Warranties

McGraw-Hill Logistics Series
Jim Jones, Editor

BARNES • *Logistics Support Training*
BIEDENBENDER • *The ILS Manager's LSA Toolkit*
HILL • *Product Support Services and Training*
LACY • *Systems Engineering Management*
JONES • *Integrated Logistics Support Handbook*
JONES • *Logistics Support Analysis Handbook*
ORSBURN • *Spares Management Handbook*
SAYLOR • *TQM Field Manual*

Warranties

Planning, Analysis, and Implementation

James R. Brennan

Principal Consultant, Product Assurance Analysts

McGraw-Hill, Inc.

New York San Francisco Washington, D.C. Auckland Bogotá
Caracas Lisbon London Madrid Mexico City Milan
Montreal New Delhi San Juan Singapore
Sydney Tokyo Toronto

Library of Congress Cataloging-in-Publication Data

Brennan, James R.
Warranties : planning, analysis, and implementation / James R. Brennan.
p. cm.
Includes index.
ISBN 0-07-007567-0
1. Customer service—United States. 2. Warranty—United States. I. Title.
HF5415.5.B74 1994
658.8′12—dc20 94-4631
CIP

2 3 4 5 6 7 8 9 0 DOC/DOC 9 0 9 8 7 6 5

ISBN 0-07-007567-0

The sponsoring editor for this book was Larry Hager, the editing supervisor was Caroline R. Levine, and the production supervisor was Donald F. Schmidt. It was set in Century Schoolbook by Carol Woolverton Studio in cooperation with Warren Publishing Services.

Printed and bound by R. R. Donnelley & Sons Company.

This book is dedicated to my Heavenly Father. Without His wisdom and strength, it could not have been written.

The inspiration, dedication, concentration, and endurance provided by His presence sustained me through the many times of discouragement and fatigue.

This is truly His book and, because of that, to God be the glory!

Contents

Preface xi
Acknowledgments xiii
Introduction xv

Chapter 1. Foundations of Warranties 1

Warranty Assessment 1
Background 2
Definition 3
Criteria of a Good Warranty 4
The Measure of Goodness 5
Consumer, Commercial, and Government Warranties 5
Warranties in Merchandising 8
Profit/Loss Implications of Warranties 8

Chapter 2. Warranty Principles 11

Realities of Warranties 11
Warranty Functions 13
Assurance versus Incentive Warranties 14
Defects versus Failures 16
Types of Warranty Policies 16
Warranty Coverages 21
Warranty Remedies 21
Risk Issues 23
Reliability Considerations 26
The Warranty Process in the System Life Cycle 32

Chapter 3. Warranty Requirements 35

General Considerations 35
Express and Implied Warranties 35
Warranty Legislation 36
Warranty Structure 41
Examples of Representative Warranty Clauses 45

Contractor Strategy 53
Importance of Warranty Requirements 55

Chapter 4. Fixed-Price Extended-Duration Repair Warranty: Reliability Improvement Warranty 57

General 57
Definition and Scope 58
Potential Benefits 59
Potential Risks 62
Potential Problem Areas 65
Application Criteria 66
RIW Case Study 67

Chapter 5. Performance Guarantees 69

Definition 69
Types 69
Advantages and Disadvantages 78
Criteria for Use 79
Examples 79

Chapter 6. Warranty Tradeoffs 87

General Considerations 87
Warranty Analysis Considerations 87
Tradeoff Issues 89
Tradeoff Process 95
Impact on Warranty Cost 98
Tradeoff Example 99

Chapter 7. Warranty Costing 105

Introduction 105
Warranty Analysis Process 105
Critical Costing Considerations 107
Warranty Cost Components 110
Benefits of Bottom-Up Costing 114
Warranty Quotation Process 114
Establishment of Warranty Price (Quote) 127
Cautions 127

Chapter 8. Warranty Negotiating 133

General Considerations 133
Negotiation Process 133
Application 136
Cautions 137
Importance of Negotiations 137
Example 138

Chapter 9. Warranty Cost Effectiveness 143

General Considerations 143
Assessing the Value of Warranties 143
Requirements 144
Cost-Effectiveness Analysis Concept 146
Cost-Effectiveness Analysis Process 154
Warranty Waivers 162
Example 163

Chapter 10. Warranty Implementation and Administration 167

Staffing 167
Asset Tracking 168
Database Management 172
Repair Considerations 175
Reporting 180
Lessons Learned 182

Chapter 11. Warranty Development 185

General Considerations 185
Government Development Activities 185
Contractor Development Activities 200
Joint Army-Industry Warranty Working Group (JAIWWG) 205

Chapter 12. The Future of Warranties 209

Global Economic Outlook 209
Total Quality Emphasis 209
Survival Strategy 210
International Warranties 211
Bibliography 213

Appendix A. Magnuson-Moss Warranty—Federal Trade Commission Improvement Act 215

Appendix B. Title 10, Sec. 2403 United States Code 229

Appendix C. Defense Federal Acquisition Regulation Supplement Subpart 246.7 233

Appendix D. Army Regulation 700-139 241

Appendix E. Secretary of the Navy Instruction 4330.17 271

Appendix F. Air Force Regulation 70-11 275

Appendix G. Marine Corps Order 4105.2 301

Appendix H. Government Accounting Office Report on DOD Warranties 323

Appendix I. Warranty Checklists 327

Index 337

Preface

This book was written to clarify the meaning of warranties and provide guidance as to the planning, analysis, and implementation of effective warranties in the consumer, commercial, and government business sectors.

It has become obvious, particularly in the government sector, that warranties have been poorly conceived and implemented. In some cases the failure to properly plan and administer the warranties has resulted in a bum rap for warranties—"warranties don't work."

In my judgment, the time has come to provide insight into the attributes of warranties and their implementation to help preclude improper conception and application of warranties. This book is dedicated to just such a result.

It is my hope that this reference book will provide practical insight to suppliers and customers, resulting in benefits to both—win-win regarding warranties.

James R. Brennan

Acknowledgments

The author is indebted to many sources of material for this book. Extensive use was made of the *Warranty Guidebook* from the Defense Systems Management College. Special thanks go to DSMC and MKI Incorporated, the writers of the *Guidebook,* specifically Bob Fout and Marcy Kester. I would like to thank the Society of Automotive Engineers (SAE) for use of material from the *Reliability, Maintainability, and Supportability Guidebook.* The specific contributors whose material was particularly helpful include Don Isaacson and Selina Reid of Texas Instruments, Inc., Jennifer Lenox of ARINC, Donna Kehrt of the Air Force, and Toby Berke of Eastman Kodak Company. I would like also to thank Sherman Burton of Burton and Associates for his preparation of Chap. 10, on warranty implementation and administration.

While not drawing information directly from the group, thanks go to the Joint Army-Industry Warranty Working Group (JAIWWG) for the informative discussions on warranty issues during the course of our work. Jim Lacy of Jim Lacy Consulting was especially helpful in getting me interested in the book project, and I am grateful to him. Last, thanks go to Larry Hager, editor, and McGraw-Hill, Inc. for giving me the opportunity to write the book and have it published.

Introduction

Warranties have become a vital part of product marketing in the consumer, commercial, and government sectors. Additionally, in all three sectors, warranties are driven by law.

The high profit/loss potential with warranties makes it imperative that the supplier and customer, regardless of the business sector, become informed on the methodology and management of warranties, highlighting the benefits and risks to both sides.

Suppliers and customers are seeking guidance to help make warranties mutually beneficial: suppliers in terms of profits and survival in a tight economy, and customers in terms of satisfaction with the products.

Chapters 1 and 2 describe warranty principles as a foundation for further discussion. Chapter 3 discusses warranty statutes across all three business sectors and examines existing warranty statements relative to requirements. Chapters 4 and 5 explain the concept and application of repair warranties and performance guarantees, respectively. Chapters 6 and 7 describe the tradeoffs of warranty requirements and the methodology of warranty costing, respectively. Chapter 8 develops the process of warranty negotiating between suppliers and customers—usually present in government and commercial sectors.

Particularly in the government sector, the cost effectiveness of the negotiated warranties must be evaluated and either a go-ahead or a waiver decision reached. Chapter 9 develops the cost-effectiveness analysis methodology to arrive at the decision. Chapter 10 discusses the implementation and administration of the negotiated warranties—a vital effort if warranties are to have value for the customer. Chapter 11 addresses the development process for warranties, including warranty planning and execution and a managers' warranty checklist. The future of warranties is examined in Chap. 12, including what it will take for suppliers to survive and the associated impact on warranties.

In Chaps. 1, 2, 3, and 12 the consumer, commercial, and government sector warranties are addressed, since they are uniquely informative. In Chaps. 5, 6, 7, 8, 10, and 11, due to redundancy, only the sector in which the subject is most thoroughly applied is addressd. In Chaps. 4 and 9 the subject is germane to only the government sector. Due to more specific requirements, government warranties are best represented by military or weapon system warranties.

In summary, this book is intended to provide a thorough, credible, and practical reference on the subject of warranties for use by both suppliers and customers. The goal of the discussion, including planning and execution, is the realization of warranties as effective business tools.

Warranties

Chapter

1

Foundations of Warranties

Warranty Assessment

As you begin reading this book, it might be interesting to personally assess your current perceptions of warranties from your experience. The vehicle for this assessment is a series of questions for you to answer. At this point in time the concern is not with giving the "right" answer but, rather, stating your answer on the basis of your opinion. The author requests that you revisit these questions while reading the book to see if the material presented in the book altered or substantiated your opinion and hopes that the book will fine-tune your knowledge such that your resulting perception of warranties will benefit you as both a customer and a supplier. The questions are

1. Should warranties be free to customers?
2. Should suppliers make a profit on warranties?
3. Should warranties be a marketing issue only?
4. Should both supplier and customer assume risk on warranties?
5. Are warranties linked to customer satisfaction?
6. Does a 7-year warranty on an item mean that the item is better than a similar one with only a 5-year warranty?
7. Do all warranties have built-in incentives for suppliers to do better than their commitments?
8. Does a warranty of a specific duration imply that the item will not fail in that duration?

9. Is a 7-year, 70,000-mi warranty on an automobile power train better than a 3-year, 36,000-mi bumper-to-bumper warranty?
10. If suppliers did not offer a warranty on their products, would you buy their products?

Additionally, as an overlay to these specific questions, consider your definition of a "good" warranty. On reading the book, revisit your definition to see if it has been altered or substantiated.

Background

> Money back guaranties [sic] or the good housekeeping seal of approval is of damn little comfort to the rifleman on the battlefield when his rifle jams. It is of little comfort to the commander who has a five-year or fifty-thousand-mile warranty on a truck if it breaks down on the way to the front with ammunition that may be critical to winning the battle.[1]

This quote from former USMC Commandant General Paul X. Kelly can best be summarized by saying that warranties don't win wars. It is important at the outset that we put warranties in proper perspective. In and of themselves warranties have no magic in improving a product's performance or quality. They are good marketing tools and may provide monetary incentives for the supplier to improve the item's quality and performance. They also provide "insurance" for the customer that the suppliers will stand behind their products, but they do not ensure perfection.

For some reason, the average consumer assumes that, if a 36-month warranty exists on an item, the item should not fail in the 36-month period. In fact, the item may fail in that period, and usually the supplier has made no claim that the item would not fail in the warranty period. Additionally, the supplier has made economic provisions for the service on the basis of an expected number of failures in the warranty period. This means, in essence, that if the item fails for legitimate reasons during the warranty period, the supplier will correct the problem for the customer at no additional cost—correcting the problem being a replacement of the item or repair of that item.

Another misconception regarding warranties involves the supposition that a particular length of the warranty relates to the quality of the product. If I were to go to a furniture store and examine the warranties on two different mattresses preliminary to purchasing the item, I might find that one of the mattresses has a 12-year warranty and another of comparable size and construction has a 15-year warranty. Is the 15-year warranted mattress "better" than the 12-year warranted mattress? Not necessarily. The 15-year warranted mattress probably costs more, and may be a better mattress, but unless the 15-

year mattress had superior construction (parts and/or labor), it is not a better mattress.

If one listens carefully to advertisements, one will hear ads that imply a longer warranty means a "better" product—not necessarily. Unless the supplier expended additional material or labor in the item's manufacture, including testing, the longer warranted item is not necessarily a better product.

In a sense, a warranty is an insurance policy for the customer. Having the warranty policy in hand, the customer can be assured that a legitimate defect or failure will be righted by the supplier, however many times the customer returns the item during the particular warranty period.

Definition

What, then, is the definition of warranty? Figure 1.1 depicts Webster's rendition[2] of a warranty and a guarantee. The warranty is actually the documentation of the guarantee to include the responsibilities of the maker if the guarantee is breached. As Fig. 1.1 shows, we will use the terms *warranty* and *guarantee* interchangeably in this book.

A warranty, then, is assurance that the supplier of an item will back the quality of the item in terms of correcting any legitimate problems with the item at no additional cost for a particular period of time or use.

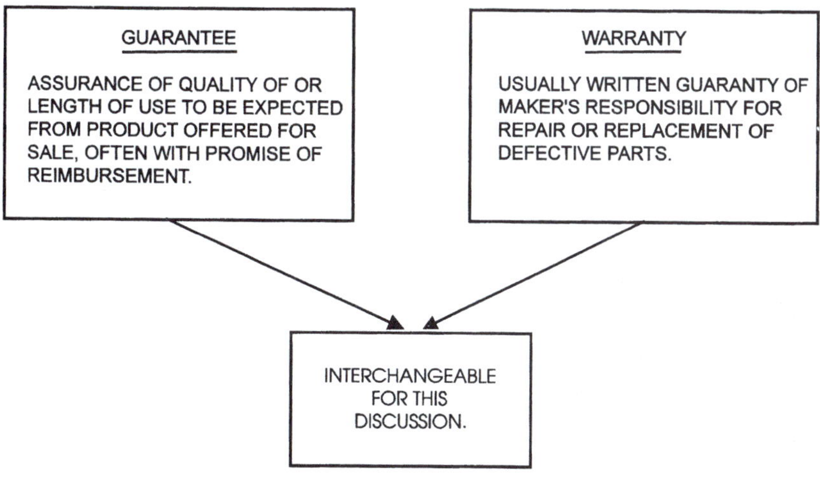

Figure 1.1 Warranty and guarantee definition.

Criteria of a Good Warranty

The question arises as to what makes a warranty good. When asked this question in a warranty symposium, the answers were quite diverse. To one person a good warranty is one related to a product that never had to be returned for repair or replacement during the warranty period. To another person a warranty is not good unless the expense of repair or replacement of the warranted product over the warranty period exceeded the amount added by the seller to the sale price of the product to cover expected warranty expenses.

In other words, the first person was not concerned about the initial cost of the warranty as long as it was never, or rarely, necessary to bring the product in to the seller for repair or replacement. As in the case of some automobiles, perhaps as much as a third of the total price of the car is warranty cost coverage. The first person is essentially saying that the peace of mind of not having to take the car in for repair or replacement is worth the cost. A key point is that the car had very few failures because of performance and quality designed and manufactured in and not because of a piece of paper offering the warranty. If the producer of the car was encouraged by management to beat the warranty stipend in the sale price as a cost-saving strategy, then and only then could the warranty be related to a more reliable, better-performance car.

The second person was not concerned about the number of times it was necessary to return the product during the warranty period as long as the "free" repair or replacement cost exceeded what that individual initially paid for the warranty. (In this and similar contexts throughout this book, the term *free* means services or expenses included in or provided by, i.e., not exceeding those specified in, the warranty.) The second person is essentially saying that the trouble and stress associated with returning a warranted item is acceptable if the dollar difference favors the free maintenance.

Most of us, with our opinions on the goodness of warranties, would probably fall somewhere in between the positions described above. The average response would probably be closer to the opinion of the first person than that of the second. Most of us are willing to bring an item back a few times during the warranty period, knowing that we paid for the warranty already in the sale price of the item. Likewise, very few of us would keep sufficiently accurate records to know when the free returns dollars exceeded the price of the warranty and would tire of frequent trips to the supplier for warranty repair or replacement efforts.

Having gone through this rationale, we will define a good warranty as one providing benefits to both the supplier and the customer—a win-win situation. The supplier has the opportunity to make a profit and therefore survive in the tight global economy. The customer has peace of mind and a higher degree of satisfaction because the equipment is serviced directly by the seller.

The Measure of Goodness

In the previous section a good warranty is defined as one affording a win-win situation for the supplier and the customer. Let us now focus on how the customer measures "goodness."

For person 2, the measure of goodness is strictly economic—dollar-by-dollar comparison. For person 1, the measure of goodness is strictly the quality of the item in terms of rarely requiring service. We will summarize person 1's measure of goodness as effectiveness. We will likewise summarize person 2's measure of goodness as cost. Since it was earlier assumed that most people would place themselves somewhere between persons 1 and 2 relative to the assessment of warranty goodness, we will then define the measure of goodness as the *cost effectiveness* of the warranty.

It is interesting to note that purist person 1 and purist person 2 could also use cost effectiveness as a figure of warranty goodness. For person 1 the effectiveness contribution is the only contributor to cost effectiveness—cost has a zero contribution. For person 2 the cost contribution is the sole contributor to cost effectiveness—effectiveness has a zero contribution. In reality, the cost-effectiveness figure of merit for warranty goodness is a composite of cost and effectiveness. A more thorough treatment of warranty cost-effectiveness analysis is provided in Chap. 9.

Consumer, Commercial, and Government Warranties

A consumer product is any tangible personal property which is distributed in commerce and which is normally used for personal, family, or household purposes. Consumer warranties are the same for all purchasers. Consumer warranties are governed by the Magnuson-Moss Federal Trade Commission Improvement Act and the Uniform Commercial Code enacted in each state. These statutes will be discussed in detail in Chap. 3. Hair dryers, washing machines, and automobiles are examples of consumer products.

Commercial products are any goods sold between merchants. Commercial warranties need not be the same for all purchasers, and, subsequent to any negotiations, the individual agreements become part of the purchase agreement. Commercial warranties are governed by the Uniform Commercial Code enacted in each state. Jet airplanes produced by aircraft companies and bought by airlines are an example of a commercial product.

Government warranties cover a wide range of applications. For our purposes we will limit government warranties to the U.S. Department of Defense (DOD) or military sector. These warranties are known as *weapon system warranties*. The Department of Transportation (DOT), particularly the Federal Aviation Administration (FAA), also uses warranties, as does the Department of Energy (DOE). The weapon system warranties afford a more informative vehicle for discussion when compared to the consumer warranties and, to some extent, the commercial warranties. Weapon systems are those materials required to accomplish military missions in the defense of our country. Weapon system warranties are governed by Sec. 2403 of Title 10 of the U.S. Code and will be discussed in detail in Chap. 3. It should be noted that the weapon systems must have a unit production cost of greater than $100,000 or a total production cost of at least $10,000,000 to be governed by the statute. Bombers, missiles, tanks, ships, radars, and communication equipment are examples of weapon systems or supporting subsystems of weapon systems.

Consumer versus Commercial Warranties

If a substantial percentage (greater than 10 percent) of the sales of commercial products is in the consumer sector, the product becomes classified as a consumer product and is subject to the Magnuson-Moss Federal Trade Commission Improvement Act. Thus, some relatively sophisticated items such as computer printers and private, single-engine aircraft have been shifted into the consumer sector and therefore become subject to federal legislation.

Full consumer warranties under Magnuson-Moss place many obligations on the seller including, but not limited to, providing remedies for claims within a reasonable time at no charge and refunding the purchase price less reasonable depreciation if the defect cannot be remedied within a reasonable number of attempts.

Commercial product warranties under the Uniform Commercial Code are generally not as restrictive to the seller as those under Magnuson-Moss.

Consumer versus Government Warranties

In the consumer business sector, when a warranted item is purchased, the sale price has been increased to cover the expected cost to the seller of replacing or repairing the item(s) for the warranty period. A warranty contract may also be purchased to cover repairs or replacement for a specific period of time beyond the original warranty period.

In the government sector the warranty cost cannot be included in the sale price of the item, but rather is costed and negotiated as a separate line item.

There are similarities in requirements between consumer and military warranties. Both warranties define what is to be warranted, the length of time the warranty is in effect, acceptable use conditions and environments, the extent of warranty coverage, and the supplier's remedies under the warranty. Specific warranty statements from both sectors will be presented in Chap. 3.

Relative to the differences in overall warranty contracting between the two sectors, several criteria can be compared as shown in Table 1.1.

The requirements for consumer warranties are defined by competitively self-determined marketing considerations. The requirements of weapon system warranties are specified by the government. Consumer warranties are benefited by extensive market research; weapon system warranties are not. Consumer items are manufactured prior to the sale; weapon systems are manufactured after the sale. Consumer warranties are generally provided in lieu of other rights and entitlements of the customer; weapon system warranties are generally provided in addition to other government rights and entitlements.

Consumer warranties enjoy utility by spreading small risk increments across massive numbers of consumers; weapon system warranties cannot spread incremental risks beyond one massive customer (government). Consumer warranties routinely employ "factory-authorized service," but weapon system warranties generally involve service performed by the user (organic maintenance). Finally, consumer warranties are associated with an "orderly" user environment, while weapon system warranties are attendant to a "hectic" user environment.

Of special interest is the subject of requirements. The weapon system supplier must adhere to a strict set of requirements (a law), while the consumer supplier's requirements are more loosely governed by law and strongly influenced by competition. The specific statutes will

TABLE 1.1 Commercial versus Military Warranties

Commercial	Military
Requirements self-determined	Requirements customer specified
Extensive market research	Limited or no market research
Manufacture prior to sale	Manufacture after sale
Factory-authorized service	Services performed by user
"Orderly" user environment	"Hectic" user environment

be covered in Chap. 3. A more detailed description of the three types of warranties can be found in Chap. 2.

Warranties in Merchandising

Warranties have long been a vital part of merchandising in the consumer sector and are becoming so in the commercial and government sectors as well.

Customers expect a warranty on the items they buy, and suppliers should be willing to offer them as a sign of the confidence they have in the overall quality of their products. As discussed earlier in this chapter, a longer warranty for one product, when compared to that of another, does not necessarily ensure a higher-quality product, but it does suggest more supplier confidence in the former product over the latter. That extra confidence may well be the factor that clinches the sale, all other factors such as price, appearance, and horsepower rating being essentially equal. In fact, the price could be somewhat higher for the longer warranty product and still be a better value in the eyes of the customer because of the longer warranty.

Unquestionably, the length and coverage of the warranty are features that impact the customer's decision to buy. These findings in the consumer sector are as valid as those in the commercial and government sectors, even though the processes are somewhat different. Implicit in any discussion of consumer, commercial, or government warranties is the goal of maximizing customer satisfaction, however that is perceived. While customer satisfaction cannot always be quantified, it is a very real factor in the merchantability of products. Ultimately, a satisfied customer should be the goal of any supplier, and the warranty is but one piece of the customer satisfaction puzzle.

Profit/Loss Implications of Warranties

For purposes of discussion, we will use the automobile as our product. Obviously, the goal of General Motors, Ford, or Chrysler is to sell cars. We have already established warranties as a key sales factor. These companies can consider any of several options. They can offer a longer, more thorough coverage "super-" warranty than the competitors' and increase the sale price of the car to cover the expected warranty cost. They can offer a "standard" warranty and increase the sale price commensurate with the expected cost with the standard warranty. They can offer the standard warranty with no increase in the sale price or the superwarranty with no increase in the sale price. Note that no or reduced warranties are not viable options in today's marketplace.

What we have, then, is a calculated gamble for that company. If their market research indicates that most consumers are willing to pay more for a car with a better warranty, then they would choose the superwarranty with the associated higher car price. If, however, the market research indicated that price was the key factor, then they would likely choose the standard warranty with the lower attendant car price. That certainly is easy enough. But what if the market research produces a muddle relative to price versus warranty? Now we are in the land of reality. In most cases, the results of the market research will be inconclusive, and we are back to the calculated gamble. The best gambler may well come out ahead. Clearly, if the warranty offered is not covered in the sale price of the car, the company can experience a loss on the warranty, but they did sell the car. So they make a profit on the sale of the car but could experience a loss on the warranty. Can they continue this way? Probably not.

Stories are told about the dilemma Chrysler encountered when, for marketing reasons, they decided to offer a 7-year, 70,000-mi warranty on their power trains. There simply was not sufficient data to indicate what they would be expected to pay in warranty coverage with this plan. They agonized over the problem. Then Chairman Lee Iacocca finally decided that they needed a marketing edge and opted to offer this extended warranty in lieu of good backup data. The rest is history. Mr. Iacocca took a calculated gamble and won.

An interesting question arises. Is the 7-year, 70,000-mi warranty on the automobile power train better than a 3-year, 36,000-mi warranty on the entire bumper-to-bumper automobile? The answer is: That depends on the customer. The bottom line is this: Which will likely sell the most cars?

Let us use my own experience as a consumer to illustrate the point. In buying my last car, I was strongly influenced by the length of the warranty—the longest available—and minimized the coverage issue. I was very pleased with myself for the first 15,000 mi. The car was performing well and was quite comfortable to drive. Disaster struck at the 15,000-mi check. Having no apparent problems up to that point in time, I was confident that the 15,000-mi service would be quite standard and therefore not expensive.

I could not have been more wrong. On calling the shop as to when I could pick up the car, I casually asked for the cost of the service and learned that it would be around $1100. When I recovered, I asked the service advisor why the cost was so excessive. He then told me that my valve cover was leaking oil, my power steering needed repair and that required removing the rack and pinion, and the air-conditioning unit was installed off-center on the axis and would therefore require replacement of the compressor and other assorted items.

I promptly informed him that the problems he described were material and workmanship-type problems and were "in there" from the start and therefore had nothing to do with wear related to mileage. He promptly informed me that I was correct, but that the material and workmanship warranty had expired. I then asked the ill-fated question: Aren't any of those problems covered under the power-train warranty?

He acknowledged that the engine oil leak via the valve cover problem could be covered. He then told me that this had been a chronic problem with this type of engine. I was getting encouraged by then. My encouragement turned to dismay when on looking it up, he informed me that my year car was not one of those covered under this chronic engine defect. I asked him if they were 2.5-liter engines, and he said they were. I asked him if mine was a 2.5-liter engine. He said it was. The next question is obvious: Why isn't mine covered, too, since they had obviously not corrected the chronic engine leaking defect? A spirited discussion ensued, and my final remark was a threatened call to the vice president of quality. The result was a $700 adjustment to the bill.

The major point of this sad tale is that at the prospect of paying $1100 for a routine maintenance check, my preference was for a 3-year, 36,000-mi bumper-to-bumper warranty rather than the "apparently better" 7-year, 70,000-mi power-train warranty.

I had not thoroughly considered the warranty impact on my decision to buy. I had failed to consider the coverage, not just the duration of the warranty in my decision. You can bet the next time I buy a car I will consider coverage, duration, and cost of the warranty in my decision.

This brings me back to the original issue. The impact of the warranty coverage on the decision to buy depends on the particular buyer. A supplier who takes a chance on offering a longer or better coverage warranty to sell more cars, despite having increased the sale price to compensate, may lose money in terms of dissatisfied customers who may never again buy a car from that company. Warranties have significant profit and loss implications.

On a brighter note, because my car has performed quite well with normal maintenance costs since the described debacle, the cost-effectiveness measure has improved and I will definitely consider buying another of that company's cars. Oh, yes, I would also do my warranty homework before the sale.

References

1. Texas Instruments, Inc., *Manager's Guide to 1985 Warranty Legislation,* 3d ed., Texas Instruments Inc., Dallas, 1990.
2. *Webster's New Collegiate Dictionary,* 5th ed., Merriam, Springield, Mass., 1977.

Chapter

2

Warranty Principles

Realities of Warranties

Following are four realities of warranties which require discussion as the foundation for further development of the subject:

- Items under warranty may fail.
- Warranties are not free.
- Warranties do not ensure particular quality and/or performance levels.
- Warranties indicate quality and/or performance levels for which the supplier accepts liability.

Somewhere in our backgrounds we have the notion that a warranty on an item for a particular length of time carries with it the assurance that the item will not fail within that time frame. Nothing could be further from the truth. In virtually every instance, the supplier does not expect the item to survive the warranty period without a failure. In fact, in many cases, the supplier makes no claim that the item will not fail.

You can examine the warranty statement on a typical consumer product and you will not find any claim of a failure-free warranty. In the government sector, particularly for weapon system warranties, a "failure-free" warranty does not mean that the item will not fail. Rather, it means that if it fails, the supplier will take care of the problem at no additional cost to the government. From the government's financial perspective, the item has not failed.

For the supplier to offer a truly failure-free warranty, they would have to spend more money in the development and testing of the item. This additional development and testing cost would be added to the sale price of the item to cover the supplier in case of failure. The result would likely be that the supplier's product would not be price-competitive.

Many consumers are not aware that the warranty they get on an item is not free. In some cases, suppliers will even advertise a free warranty. Just recently, I heard an advertisement on the radio where a car dealer noted that he would offer the 3-year, 36,000-mi warranty free on said new automobiles for sale. The fact is that in the commercial market, the expected cost to the supplier in fulfilling the warranty agreement is included in the price of the item—there is no free lunch. Suppliers are in business to make a profit. It would not take too many truly free warranty problems to bankrupt a supplier.

In the government there has been an effort to emulate commercial warranties with respect to weapon system warranties. To date, this has been a dismal failure for two reasons: (1) weapon systems are much too complex for simplistic commercial warranty requirements, and (2) the government has been attempting to acquire no cost warranties "like the commercial customers do." Surprise: As we have just firmly established, warranties are not free in the commercial market. What actually happened is that the weapon system suppliers were not willing to offer no-cost warranties, so they buried the cost in other contractual areas such as the quality and reliability programs. The government has made a commendable adjustment in recent months to allow costs for warranties.

Let's examine why government suppliers are not willing to offer no-cost warranties. First, the warranty is a separate line item in virtually every government contract. Under a warranty, a supplier is required to be the depot for repairs of their equipment; this involves an added expense since the basic contract calls for the government to be the depot. Associated with the contractor repairs are administration costs for records, reports, and review board meetings. Additionally, contractors are required to make not-to-exceed fixed-price quotes on items long before they are built and tested. There is therefore supplier risk introduced which also should be priced to cover the contractor for unforeseen design and material and workmanship problems.

The argument against no-cost warranties in the consumer area is much simpler. Suppliers are in business to make a profit in all facets of their business. They can ill afford excessive repair costs which have not already been included in the price of the item, not to mention chronic design problems.

As has been stated previously, the existence of a piece of paper called a "warranty agreement" does not magically provide iron-clad assurance that the item will perform any better or run more reliably than without the warranty. The improved performance or better reliability comes only through better design and production effort on the item.

If warranties don't ensure better performance and better reliability, then what good are they? Warranties ensure that suppliers accept liability for the levels of performance and quality they are offering. The liability involves correcting or repairing any breaches to what they are offering. A supplier who gets the money up front has the incentive to produce higher-performance and higher-quality items to keep as much of the money as possible. This incentive might be lucrative enough to the supplier to design out and produce out problems before they get to the consumer and therefore reduce the number of warranty breaches. Result: a possibly more satisfied customer.

Warranty Functions

Warranties are contractual instruments. They are tools used by the customer in the acquisition process to help obtain systems with improved quality and acceptable life-cycle costs. The following is a classification of warranty functions:

Assurance validation

Warranties help assure the customer that the supplier delivers a product whose design and manufacture and materials and workmanship conform to contractual specifications. Since it is assumed that such defects can be avoided by careful and prudent management during the development and production of the warranted item, the costs of providing remedial action should be borne by the supplier. Assurance validation ends at system acceptance for design defects and after a reasonable period of time with respect to material and workmanship defects.

Incentivization

Warranties, by their employment, incentivize the supplier to some degree. The incentivization function, however, becomes distinctive when the guarantee provisions define penalties for failure to achieve guaranteed performance parameters or rewards for exceeding such parameter levels.

Insurance

Every warranty provides some insurance against the risks of repair and replacement costs to the customer. This risk can be quite significant in the initial postacceptance years.

Assurance versus Incentive Warranties

Assurance warranty

When the primary intent of the customer is to ensure that minimum design, quality, and performance levels are achieved, assurance warranties are appropriate. The customer is not seeking anything more than what the contract specifies, and the terms and conditions of the warranty do not provide any incentives for the contractor to do any better.

Incentive warranty

When the intent of the customer is to encourage the supplier to exceed minimum design, quality, or performance levels, incentive warranties are appropriate. For this type of warranty, the supplier can opt to merely meet the requirements or can seize the opportunity to achieve improved performance and quality to keep more of the money, and hence the incentive. The requirement should be written so that the profit potential to the supplier is sufficient to induce the contractor to exceed minimum levels.

Comparison of assurance and incentive warranties

Table 2.1 compares assurance and incentive warranties with respect to various factors. To further describe the distinction between assurance and incentive warranties, the following example is presented.

Expected field mean time between failures (MTBF) of warranted item = 1000 operating hours

Expected operating hours in warranty period = 200,000 h (all systems)

$$\text{Total number of expected failures} = \frac{200{,}000}{1{,}000} = 200 \text{ failures}$$

For the assurance warranty, the terms and conditions may state that the customer will repair or replace all failures that are expected when the MTBF is met (maximum of 200 failures) and the contractor will be responsible for all failures beyond 200 that occur during the warranty period, at no additional cost to the government. The contractor, therefore, does not benefit from producing systems with better than a 1000-h MTBF.

Considering the incentive warranty for the same example, the terms and conditions now state that the contractor is to provide depot repair services for this system over the warranty period at a fixed price. This fixed price is based on the required 1000-h MTBF, which translates to

TABLE 2.1 Assurance versus Incentive Warranties

Factor	Assurance	Incentive
Intent	Meet minimal performance	Exceed minimum performance
Price	May be minimal	May be significant
Duration	Limited: ≤ 2 years	Extensive: ≥ 3 years
Administration	Generally moderate	May be complex
Contractor	Limited control and improvement	Significant control and improvement
Competition	May sustain competition	May reduce competition
Technology	Within state of the art	Pushes state of the art

200 expected failures. The contractor desires to keep as much of the up-front money as possible. They clearly see that each failure that can be eliminated results in more profit. Aware of their pending warranty commitment, the contractor has the incentive to invest more heavily than normal in design, production, and quality assurance to reduce the number of future actual failures below the negotiated 200 failures. Additionally, the contractor has the incentive to identify and fix pattern or systemic problems to reduce or eliminate such problems.

Figure 2.1 plots the expected profit/loss versus the achieved MTBF for the assurance and incentive warranties per this example. For both forms of warranty, the contractor will suffer a loss for a MTBF less than $1000 - X$, where X is the decrease in MTBF from the 1000 h "covered" by the warranty profit/risk dollars in the contract price. For the assurance warranty the contractor's profit rises to the expected con-

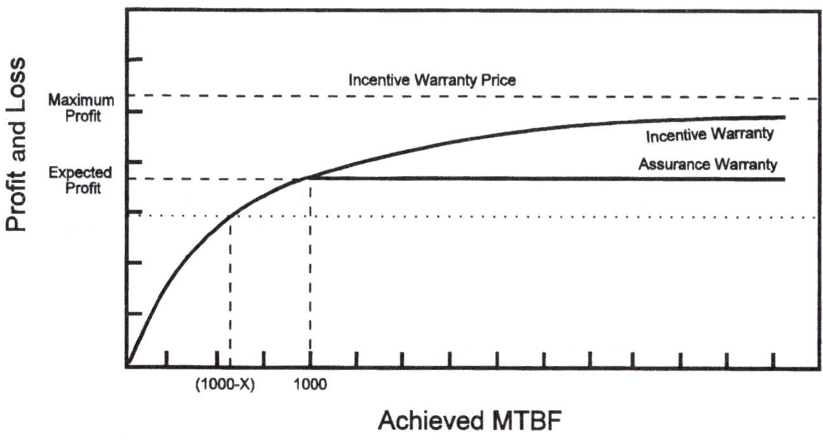

Figure 2.1 Contractor profit for assurance and incentive warranties.

tract profit and levels off for MTBF equal to or greater than 1000 h. For the incentive form, the profit continues to rise with increasing MTBF until it is asymptotic to the contractor warranty price—the costs not sensitive to MTBF, namely, warranty administration and warranty data, being the only costs incurred as MTBF approaches infinity.

With the potential for an enhanced win-win situation associated with incentive warranties, an examination of pertinent criteria for their applicability should be helpful. The following are these criteria:

- Money available for extended coverage
- Need for improved field performance
- Opportunity for improved field performance
- Contractor control of system capabilities before and during deployment
- Duration long enough to influence contractor (> 2 years)
- Plans for future competition not seriously affected
- Terms and conditions that can be written to provide adequate compliance determination and remedies

It should be pointed out that not all of these criteria may exist for any one program, and their absence does not preclude the use of incentive warranties.

Defects versus Failures

The following are definitions of these two often misunderstood terms essential for further discussion of warranties:

- *Defect.* A condition or characteristic of the warranted item(s) which is not in compliance with the requirements of the contract.
- *Failure.* Breakage, damage, or malfunction of the warranted item which renders it unserviceable for use.

It should be further noted that a failure could also be a defect, and that not all defects result in failures and not all failures result from defects.

Types of Warranty Policies

Consumer and commercial

Free replacement. The seller pays the entire cost of the remedy if the product fails before the end of the warranty period.

Pro rata. If an item fails before the end of the warranty period, the cost of the replacement depends on the age or wear of the item at the time of failure. Replacement items usually carry a new warranty. Batteries and tires are two prominent examples of items which generally carry pro rata warranties, since they exhibit wear characteristics from the beginning of use.

Table 2.2 shows a Nissan battery pro rata chart. As would be expected, if the 48-month battery fails, for example, at 24 months, Nissan Motor Company (NMC) pays 50 percent of dealer net and the customer pays 50 percent of the suggested retail price.

Combination free replacement and pro rata. An initial free replacement period is followed by a pro rata period during which the cost of the replacement item is based on a sliding scale determined by item age or wear at the time of failure during the pro rata period. This warranty would be applicable to an item which has a constant failure rate for the initial period of time and then exhibits wear characteristics (increasing failure rate) from then on.

TABLE 2.2 Nissan Battery Pro Rata Chart

Months of Service	% NMC pays of dealer net*	% customer pays of suggested retail	Months of service	% NMC pays of dealer net*	% customer pays of suggested retail
3	100.0	0	26	45.9	54.1
4	97.7	8.3	27	43.8	56.2
5	89.6	10.4	28	41.7	58.3
6	87.5	12.5	29	39.6	60.4
7	85.5	14.5	30	37.5	62.5
8	83.4	16.6	31	35.5	64.5
9	81.3	18.7	32	33.7	66.6
10	79.2	20.8	33	31.3	68.7
11	77.1	22.9	34	29.2	70.8
12	75.0	25.0	35	27.1	72.9
13	73.0	27.0	36	25.0	75.0
14	70.9	29.1	37	23.0	77.0
15	68.8	31.2	38	20.9	79.1
16	66.7	33.3	39	18.8	81.2
17	64.6	35.4	40	16.7	83.3
18	62.5	37.5	41	14.6	85.4
19	60.5	39.5	42	12.5	87.5
20	58.4	41.6	43	10.5	89.5
21	56.3	43.7	44	8.4	91.6
22	54.2	45.8	45	6.3	93.7
23	52.1	47.9	46	4.2	95.8
24	50.0	50.0	47	2.1	97.9
25	48.0	52.0	48	0	100.0

*30% parts markup payable on 100% reimbursement only.

Fleet. A large quantity or fleet of items is purchased by a single customer. The guarantee is made on the expected or mean life of the item. Compensation is often in terms of free parts or the entire item if it is nonrepairable. The keys to applicability are a large sample and one customer.

Military

Failure-free warranty. This is the most conventional type of warranty. Although it may have both assurance and incentive features, it is typically employed as an incentive warranty. The failure-free warranty is generally classified as a repair or replacement warranty designed to cover defects in material and workmanship which manifest themselves as failures of the item. In most cases, the contractor is required to fix all legitimate failures that occur during the warranty period at no additional cost to the government. The prime advantages of the failure-free warranty are simplicity and early identification of defects since the contractor has responsibility from failure 1 on. A disadvantage might be the cost associated with the higher risk assumed by the contractor.

The term *failure-free* in this context means that the government does not pay for the failure when it happens. It does not mean that the item will not actually fail, as in a failure-free burn-in test.

Threshold warranty. This warranty is also generally classified as a repair or replacement warranty designed to cover defects in material and workmanship. The threshold warranty defines a breach only when the number of failures exceeds a stated "threshold." This is a form of an assurance warranty. The threshold is typically determined by dividing the expected operating hours during the warranty period by the required field MTBF. It should be noted that this is a negotiable item. The contractor is obligated to fix only legitimate failures in excess of an allowed number, the government fixing all failures up to and including that number, at no additional cost to the government.

The threshold warranty represents a reduced risk to the contractor since there is a "grace" number of failures before remedial action is required. It is recognized that malfunctions will occur despite the best design and manufacturing processes. There is some difficulty in selecting the appropriate threshold since the required field reliability is not necessarily the same as the guaranteed field reliability.

The main disadvantage to the government is the extensive data collection, recording, and accounting that must take place regardless of whether the expected threshold is breached and tangible benefit is received.

The threshold warranty is also known as an *expected-failure warranty.*

Systemic defect. A systemic defect is one which occurs with a frequency, sameness, or pattern to indicate logical regularity. It implies a root-cause problem in the design which could cause recurring defects or failures. The systemic warranty is generally designed to cover design and manufacturing defects per the statute. When a systemic defect is uncovered, the government assumes that all warranted items produced under like design and manufacturing circumstances are similarly defective and require correction on a fleetwide basis. The systemic warranty is more apt to cover a cause rather than a symptom. All services have established quality-deficiency reporting mechanisms which need not be duplicated for this purpose—a significant advantage to the government.

Under the other warranty types, a contractor is prone to correct a malfunction with a part or assembly that will not correct the root-cause problem for the duration of the warranty. On completion of the warranty, the government inherits the problem. The remedy typically associated with a systemic warranty is redesign or retrofit.

Defect-free warranty. This warranty type directly relates to contract nonconformities rather than hardware problems. The defect-free concept is based on the assumption that nonconforming material might be delivered despite the efforts of both the government and contractor. The defect-free warranty concept underscores and preserves the contractor's obligation to deliver systems that conform to contract requirements beyond acceptance. It also recognizes that not all defects result in failure and not all failures result from defects.

A single expiration date is used for all warranted systems. This facilitates warranty administration, precluding warranty markings and documents, since the two parties only need to know the contract number to determine the applicability of the warranty. Existing systems for reporting defective material are used, therefore precluding the need for additional data systems. The reported deficiencies are processed in accordance with normal quality-deficiency report (QDR) processing procedures. Since defect-free warranties do not change what the contractor is required to deliver and the administration details are minimized by both parties, they are normally cost effective.

As with failure-free warranties relative to repair, the defect-free warranty requires that the contractor correct all defects at no additional cost to the government. If the contractor is responsible for defects only when a pattern or logical regularity is detected, then the systemic warranty shall be used, with its associated systemic remedies.

Performance guarantees. The previously discussed warranty types are concerned largely with repair of failures and elimination of defects in the design, ensuring that the contractor would be responsible for the design and material and workmanship of their system in the post-acceptance period in terms of correcting the defects and/or fixing the failures.

Performance guarantees add another dimension to the contractor's responsibility for the equipment. Performance guarantees require the contractor to demonstrate particular levels of performance after deployment for the duration of the warranty. Failure to do so makes the contractor liable for penalties such as free spares, free maintenance, dollars, and/or redesign and retrofit at no additional cost to the government. It should be noted that the warranty can be constructed to provide incentives as well as penalties if the required performance levels are exceeded.

The following are typical reliability, maintainability, and supportability (RMS) performance guarantees:

- Mean time between failures (MTBF)
- Mean time between removals (MTBR)
- Mean time to repair (MTTR)
- Turnaround time (TAT)
- Built in test (testability) (BIT)
- Logistics support cost (LSC)
- System mission readiness (availability) (SMR)

Typical operational performance guarantees include the following:

- Accuracy
- Range
- Resolution
- Speed
- Thrust
- Specific fuel consumption (SFC)

The more readily applied individual performance guarantees will be discussed in detail in Chap. 5.

Generally, RMS-type performance guarantees are preferable to the operational performance guarantees. The operational characteristics can be demonstrated one time in acceptance testing and, since they are basically not time- or durability-dependent, there is little value to recurring demonstrations after deployment of the system. RMS performance guarantees provide an ongoing indication of the system's ability to function at particular levels relative to time, wear, and maintenance activities. Additionally, data systems are already in place to collect the data necessary to verify compliance. Regardless of whether RMS or operational performance parameters are chosen for use, the requirement must clearly specify which performance parameters are to be measured, how and how often or when they are to be measured, the criteria for success or failure, and the associated penalties or incentives to the contractor. Anything short of this level of detail makes it virtually impossible for the contractor to estimate risk relative to the performance portion of the warranty.

Warranty Coverages

There are two types of coverage relative to military warranties:

- *Individual item.* Warranty coverage which requires individual warranty claim actions for each failure or defect. These claim actions will be made on a warranted item only when the item is to be repaired or replaced at the depot level. The individual-item coverage is usually associated with the failure-free and threshold warranty types.
- *Systemic.* Warranty coverage which requires a contract remedy, specifically, redesign and retrofit. The systemic coverage is usually associated with the systemic and defect-free warranty types.

It should be pointed out that both individual item and systemic coverage can be invoked on the same warranted item(s) within a contract.

Warranty Remedies

A warranty remedy is the action the contractor must take if the product does not meet the requirements stipulated in the warranty statement. The following are standard remedies associated with military warranties.

Repair and replacement

A defect or failure may be corrected through a repair or replacement action. Typically, such a remedy would be applied under individual-item coverage as opposed to systemic coverage. If the contractor performs the repair or supplies the replacement, there is no additional cost to the government. If the government performs the repair or supplies the replacement, it may bill the contractor—the term *bill back* is used to describe this remedy. The amount or the method used to determine the amount is generally specified in the contract. Normally, the bill-back amount cannot exceed the contractor's normal repair and replacement costs.

Price adjustment

In some cases, correction of a defect or failure may not be possible or practical. In these cases the only remedy available may be to "equitably" adjust the contract price downward. In this case the amount of the adjustment must be commensurate with damages suffered by the government. It should be pointed out that the term *equitable adjustment* is relative. Contractor and government may have quite different perceptions of the term. If possible, this term should be quantified or at least bounded in the warranty requirements.

Redesign and retrofit

Redesign and retrofit is a set of activities and materials to correct a design or manufacturing defect. The redesign-retrofit remedy is generally applied under systemic coverage. The set of activities and materials associated with redesign and retrofit include the following:

- Engineering analyses to determine causes of nonconformance
- Corrective engineering design and drawing changes
- Modification of warranted units and spares, as required
- Retest, retrofit, and configuration management actions

If a defect pertains to the whole population of warranted items, warranty terms may stipulate a redesign-retrofit remedy. Above all other remedies, the redesign-retrofit remedy offers assurance that deficiencies in the design or manufacturing processes will be corrected. In many cases redesign-retrofit remedies are tied to failure to meet essential performance requirement (EPR) requirements. If the redesign-retrofit remedy is not tied to EPR breach, then it is essential that some criterion be established to indicate to the contractor when the remedy is to be enforced. For example, one redesign-retrofit trigger might be when the demonstrated failure rates reach some multiple of the predicted values. Another might be when lot inspection indicates that at least a certain percent of the item are defective. Redesign and retrofit are extremely costly to the contractor and the contractor must have some criterion on which to estimate their risk relative to this remedy.

Penalties

These remedies are generally tied to the failure of the contractor's equipment to achieve the required levels of performance guarantees.

They typically involve the supply of additional assets by the contractor or extension of the warranty period or, in general, actions which cause the contractor immediate fiscal inconvenience or longer-term commitments than expected. The following are examples of penalty remedies:

- Free spares
- Free maintenance
- Extension of warranty duration
- Cost reimbursement

Relative to the free spares remedy, the warranty requirement might have been worded such that the contractor must supply a free spare for every 5 h under the MTBF guaranteed value. With respect to the extension of the warranty duration, possibly the warranty stated that the warranty on each item would be extended one day for each day the depot TAT exceeds the TAT requirement. For the cost reimbursement, the warranty could have required that for each percentage point of availability below the requirement, the contractor must pay $1000 per system. There would likely be a "floor" value to limit the contractor's liability.

Risk Issues

General

Implicit in the business of warranties is risk. Both supplier and customer assume risk in the warranty process.

The supplier's risk centers on two major concerns:

- Being covered for repair or replacement costs
- Being competitive

This situation is a paradox. If the supplier includes sufficient cost in the sale price of the warranted item to comfortably cover the expected repair or replacements costs, the sale price may not now be competitive with those of other suppliers. A supplier who shaves the expected warranty costs to keep the sale price of the item competitive with those of other suppliers may lose a significant amount of money honoring the warranty. In one case sales are risked and in the other case operating cost is risked.

It is the supplier's task to balance these risks through tradeoff analysis to arrive at the least total cost solution.

The customers' risk is concerned mainly with satisfaction with the product in terms of its cost effectiveness—will they get "value" from the warranty?

Particular

Military warranties, particularly those governed by statute (Section 2403 Title 10 U.S. Code), present unique problems since, in most cases, the contracts are fixed-price and the contractor gets paid up front. Following are risk issues for contractors relative to weapon system warranties for illustration:

- Understanding warranty requirements in the request for proposal (RFP) or request for quote (RFQ)
- MTBR versus MTBF requirements
- Maintenance concept
- Quoting fixed-price warranty with limited or no reliability data
- Possibility of situations not totally under contractor control
- Competition

It is not an easy process to write a warranty requirement that meets the intent of the law and is clear, thorough, and enforceable. Often the warranty statement prepared by the government customer is vague and incomplete. Contractors are faced with responding to the requirements without answers to their questions and the prevailing feeling that they are missing something or are underassuming the requirements. They either cover this uncertainty with excessive risk money in the quote and jeopardize their competitiveness or take a chance on losing money on the warranty if they are awarded the contract by not including risk money.

MTBR warranties, for the purpose of reducing administration problems, count every removal as a failure under the warranty. Included in these removals are those items that check OK when returned to the contractor. On the other hand, MTBF warranties count only "true" failures, namely, those that check bad (fail inspection) at the contractor's facility. MTBR warranties, which are less costly to administer for the government, represent a source of risk to the contractors since they will receive a marginally predictable number of extra returns to process that were removed for cause.

The maintenance concept is a strong driver in life-cycle cost (LCC) and warranty cost. Under the maintenance concept, the government may choose to remove and replace modules at the user level or to remove and replace line-replaceable units (LRUs) of which these modules are a part. If the government chooses to remove and replace modules rather than LRUs, the contractor incurs extra risks. The handling of modules by maintenance personnel produces more induced failures of the warranted item than handling of a larger item (LRU). These induced failures must be corrected by the contractors, and it is possible that they did not adequately predict this added cost.

Probably the single largest source of risk to the contractor is the requirement to quote a fixed-price warranty on production units long (3 to 5 years) before the units are built. The contractor must estimate the number of expected returns from reliability assessments or predictions related to the paper design. The longer the period of time between the quote and the actual production, the greater the risk to the contractor.

Often the warranty statement will require the contractor to assume responsibility for situations not directly under their control. For example, the contractor may be asked to guarantee TAT of failed units returned to the depot facility for repair and back to the customer. If the TAT includes the shipment by the customer of the unit from the field to the contractor's facility, the contractor has to guarantee that portion of the TAT that is the government's responsibility. If the government should tarry on the shipment, the contractor could be in breach of the TAT requirement.

Last, but certainly not least, is the risk issue of competition for the contractor. In a competitive procurement, the competition's presence is continually felt. The exceptions that contractors take to the warranty requirements or the amount of risk money they include in their quotes are always questioned with respect to what the competition might agree to or how much risk they are willing to take to win the contract.

Following are risk issues for the government customers with respect to weapon system warranties for illustration:

- Improved reliability and reduced support costs
- Implementation within baseline maintenance concept
- New data-collection systems
- Effectiveness of administration

It is the goal of warranties, by requiring that contractors be held accountable for the performance and repair of their units postacceptance, to improve the reliability and reduce the support cost of government procured products. If the government writes a clear, thorough, enforceable warranty requirement with adequate incentives for the contractor to improve product quality over the course of the warranty period, the government could well realize these benefits. Anything short of this preparation, however, could jeopardize these benefits. As was stated earlier, the sole existence of a warranty requirement does not ensure value to the customer. The contractor must be motivated to build a better product because of the warranty. There is certainly risk for the government customer in procuring a warranty.

The warranty must be implemented within the baseline maintenance-support concept for the warranty to be cost effective. If the maintenance and administrative personnel conduct the warranty with special maintenance procedures and a separate supply line, the economic benefits of the warranty are doomed. Without adequate procedures and training to help the warranty's implementation within that program's maintenance-support concept, it is highly likely that the warranty will result in duplication of assets and efforts.

The creation and upkeep of maintenance data-collection systems are extremely costly to the government. Likewise, without the capture and processing of valid warranty performance data, the warranty cannot be properly enforced. There is no quick and easy solution to this dilemma. One solution is the inclusion of sufficient circuitry and software in the warranted item to capture and process the required performance data. This would eliminate the use of hand-entered data present in existing field data-collection systems and hence reduce the labor and measurably improve the accuracy of the data. For warranties to be effective on the complex new and future weapon systems, this approach must be carefully traded off against the increased cost of the capability. To some observers, it is the only hope of warranties with value to the government.

The government will continue to experience shortages in maintenance and support personnel to administer warranties. In many cases, the failed units are returned to the contractor for repair well after the unit is out of warranty and, in some cases, never returned. For warranties to have value to the government, the failed or defective items must be returned to the contractor in a timely manner for correction and reissue for use. The electronic collection and processing of data previously mentioned will certainly be an assist relative to the workforce shortages, but the number and depth of warranties must be examined relative to the availability of workforce to effectively administer them. The use of incentives for maintenance and support personnel relative to the processing of warranty claims bears investigation also.

Reliability Considerations

Reliability is a significant consideration in warranty requirements and implementation. The frequency of failure, MTBF, derived from the warranted item's expected failure rate is a critical parameter in determining the expected number of failures during the warranty period for warranty pricing. Additionally, once the equipment is fielded, the number of failures experienced is essential in determining the achieved MTBF for performance guarantee determinations.

Basically, warranty reliability can be classified in two categories: constant failure rate governed by the exponential distribution, and increasing failure rate governed by the Weibull distribution. Electronic equipment reliability historically follows an exponential distribution, while mechanical or electromechanical equipment typically follows a Weibull distribution. We will discuss the two separately.

Exponential distribution

Figure 2.2 presents the curve of failure rate λ as function of time. This curve is often referred to as the "bathtub curve."

During the infant mortality portion of the equipment life, the failure rate decreases with the uncovering and elimination of inherent failures of the equipment. During the useful life portion of the curve the failure rate remains essentially constant, a desirable feature statistically, and finally during the wearout period, the failure rate is found to increase with time. Hence, once the system is burned in and before the wearout occurs is the desired period of consideration for electronic equipment.

For equipment with a constant failure rate over time

$$\text{MTBF} = \frac{1}{\Sigma\lambda}$$

where $\Sigma\lambda$ is the total failure rate (failures $\times 10^{-6}$ h) across all items in the equipment

$$R(t) = e^{-\lambda t} = e^{-(t/\theta)}$$

where λ = total equipment failure rate
θ = equipment MTBF
t = mission time
$R(t)$ = probability of completing a mission of time t without equipment failure

Thus, for electronic equipment in its useful life, the MTBF can easily be determined from the failure rates and used to calculate the expected number of failures during the warranty period. Additionally, the probability of passing a test to demonstrate a particular performance level can also be easily determined using the exponential function if only the failure rate or MTBF and the length of the test are known.

The following are important statistical facts relative to the exponential distribution:

- Electronic systems conforming to their contractual MTBF require-

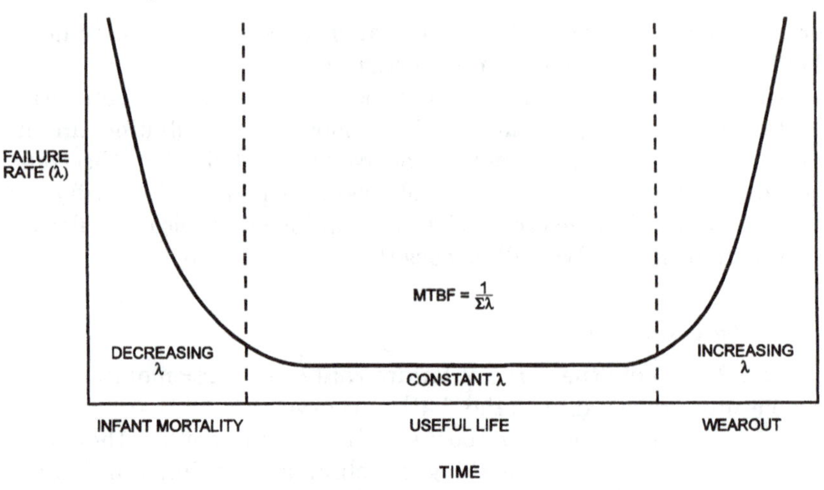

Figure 2.2 Failure rate as a function of time.

ments will fail to demonstrate that MTBF (from observed hours and failures) in over 50 percent of the tests.

- Electronic systems conforming to their contractual MTBF requirements will demonstrate only 87 percent of the required MTBF in over 90 percent of the tests.
- Uncensored field data (hours and failures) usually yields calculated field MTBFs much lower than the predicted value. In most cases the field MTBFs are one-fifth to one-half of their predicted values.
- The probability that electronic equipment that meets the contractual MTBF requirement will operate without failure for a time equal to the MTBF requirement is only 37 percent.

The first two statements seem to be in error. The reason behind the two statements lies in the statistical confidence levels associated with demonstrating MTBF from actual hours and failures. The point is that to pass these tests, the field MTBF would need to be better than the contractual MTBF. Without this knowledge, one could feel secure with a MTBF that meets the requirement and would not include sufficient risk in their warranty quote relative to meeting the MTBF performance guarantee.

The third statement is quite important relative to estimating expected failures during the warranty period. Rarely is the field MTBF as good as that predicted. Using the predicted MTBF directly without some factoring is dangerous since it would result in underestimating the number of failures to be repaired during the warranty. The concern is not with predicted MTBF since in most cases that is all that is available at the time of warranty quoting. The concern relates to use of unfactored MTBFs, therefore not accounting for induced failures due to rough handling and/or improper maintenance in the field.

In the last statement, one must be sure not to assume that since the time of the test or mission is equal to the contractual MTBF requirement, the probability of the electronic equipment operating without failure for that period is 50 percent. Here again, using 50 percent rather than the correct 37 percent could result in underestimating the true risk to the contractor in passing a test associated with a performance guarantee. It should be noted that the probability is always 37 percent regardless of the MTBF value or the mission length, as long as they are the same. Figure 2.3 shows the exponential reliability function. From the figure, the 37 percent point can be seen.

Weibull distribution

The Weibull distribution is a more complex statistical distribution than the exponential. Where the exponential distribution is a one-parameter (λ or θ) distribution, the Weibull is a three-parameter (α, β, γ) distribution. Since our intent is not a statistical discourse, we will not go into detail relative to the significance of the three parameters. Suffice it to say that β, the shaping parameter, helps us to see the relationship between the Weibull and the exponential distributions. Specifically, when

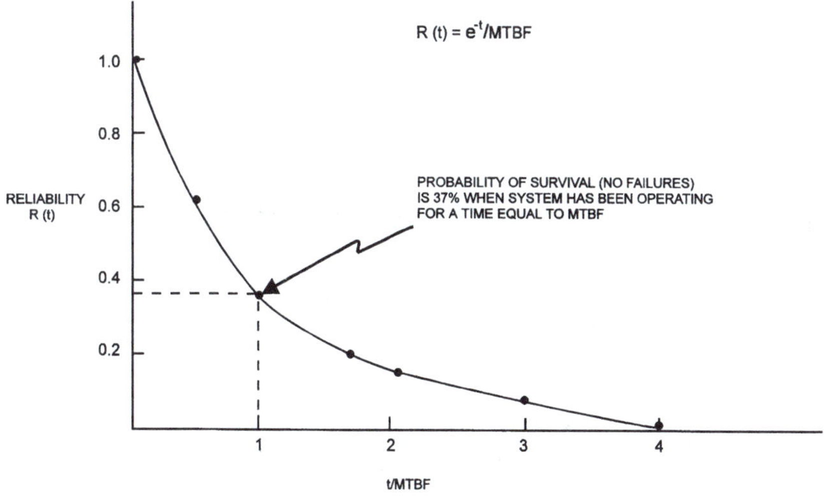

Figure 2.3 Exponential reliability function.

$\beta = 1$, the Weibull reduces to an exponential distribution. This will be illustrated later.

Figure 2.4 shows the curves of hazard rate (failure rate) as a function of time for three different values of β. Note for $\beta = 0.5$, the hazard rate is decreasing over time. For $\beta = 5.0$, the hazard rate is increasing over time. Typically, the hazard rate increases over time for items with a wearout characteristic. Such items include batteries, tires, and jet engine turbine blades. Note from Fig. 2.4 that the hazard rates for wearout items ($\beta > 1.0$) increase rapidly beyond approximately .6 multiple of θ. This information is useful for estimating the expected number of returns for pricing warranties for wearout items.

For equipment with a Weibull failure distribution over time, the hazard (failure) rate is expressed as

$$h(t) = \frac{\beta}{\theta}\left(\frac{t}{\theta}\right)^{\beta - 1}$$

where θ is the mean life of the item and t is the mission time. The mean time to failure (MTTF) is expressed as

$$\text{MTTF} = \theta\Gamma\left(\frac{1}{\beta} + 1\right)$$

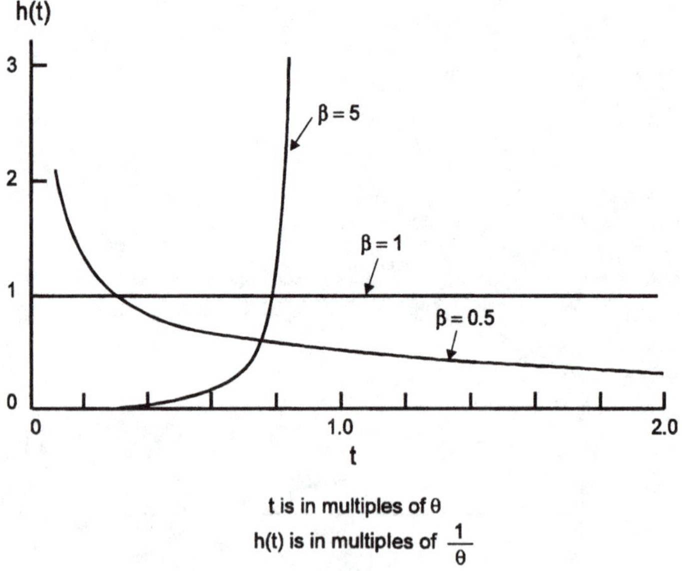

Figure 2.4 Weibull failure (hazard) rate as a function of time.

where Γ is the gamma function. Reliability is expressed as

$$R(t) = e^{-(t/\theta)^{\beta}}$$

Substituting for $\beta = 1$ in these three equations yields $h(t) = 1/\theta$, MTTF = θ, and $R(t) = e^{-(t/\theta)}$. These are the three relationships defined for the exponential distribution.

It should be pointed out that rarely does $\beta = 1.0$, and making that assumption could produce very conservative results with respect to the number of expected failures when one considers wearout items. The assumption that $\beta = 1.0$ for wearout items is much more unlikely than the assumption that electronic items are in the useful life portion of their failure rate life and can therefore use the exponential distribution. Also, it is interesting to note that MTTF, not MTBF, is more appropriate for the Weibull distribution. For virtually all wearout items, repair is not possible or feasible; hence the item fails only once before it must be discarded. In other words, there is no renewal of that item. Then, MTTF is more appropriate than MTBF.

Figure 2.5 illustrates the Weibull reliability function for the three values of β.

As Fig. 2.5 shows, the reliability for the wearout item, $\beta > 1$, is higher for times < 1.0 multiple of θ, than that for $\beta = 1.0$ and $\beta < 1.0$, but then degrades rapidly. On the contrary, the reliability for items with a de-

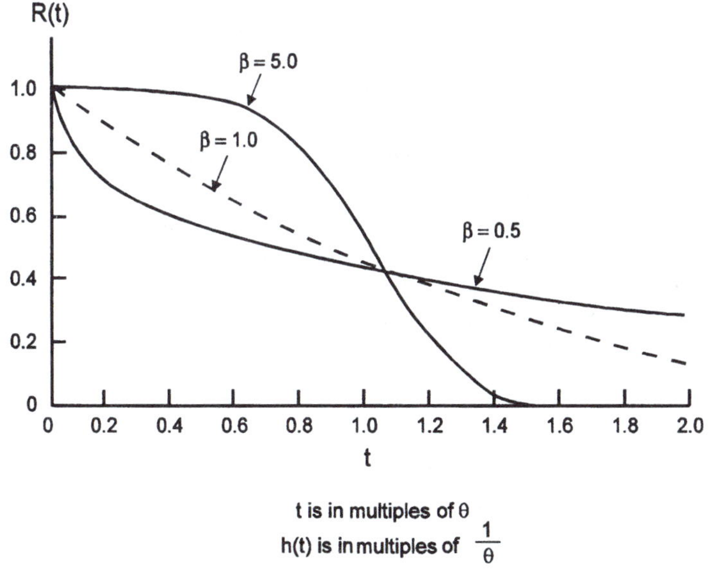

Figure 2.5 Weibull reliability function.

creasing failure rate over time is lower for times < 1.0 multiple of θ, than that for β = 1.0 and β > 1.0, but then flattens out. Of course, β = 1.0 is the exponential reliability function previously shown. Also, for all three values of β, the curves converge at t = 1.0 multiple of θ and yield a $R(t) = .37$. We now see that the statement that .37 is the probability that an item will survive without failure for a time equal to its expected time to failure is not only valid for the exponential distribution but for all Weibull distributions, regardless of the value of β.

Finally, Fig. 2.5 tells us that for wearout items, the reliability degrades rapidly, beyond approximately .6 multiple of θ. This information is important for estimating the probability of passing performance requirements, as applicable, in the warranty.

As an illustration of Weibull reliability estimation, consider a jet engine. More specifically, consider the high-pressure turbine, assuming that it has 200 blades. It is further assumed from historical data that turbine blade reliability follows a Weibull distribution (wear out with use). The reliability of the high-pressure turbine (HPT), considering the blades only, is given by

$$R_{\mathrm{HPT}}(t) = \prod_{i=1}^{200} R_B(t) \qquad \text{and} \qquad R_{\mathrm{HPT}}(t) = \prod_{i=1}^{200} e^{-(t/\theta)^{\beta}}$$

This relationship holds when the turbine blades are independent, that is, when the failure of one blade does not cause the failure of another. One can easily see the impact of the reliability of one blade on HPT reliability.

It is important to be able to classify the failure characteristics of the warranty equipment before any estimates of expected returns are made. Often a warranted item contains both constant failure rate and wearout items and both the exponential and Weibull statistical properties are needed. Not enough can be said regarding the need for historical data to adequately estimate not only the failure characteristics but also the statistical parameters such as θ and β. Remembering that warranties involve estimation, estimation based on data is always preferable to gut-feel estimation.

The Warranty Process in the System Life Cycle

It is important to understand how the warranty fits into the product's life cycle. For illustrative purposes we will look at the warranty process within the government. Figure 2.6 shows the warranty process within the system's life cycle.

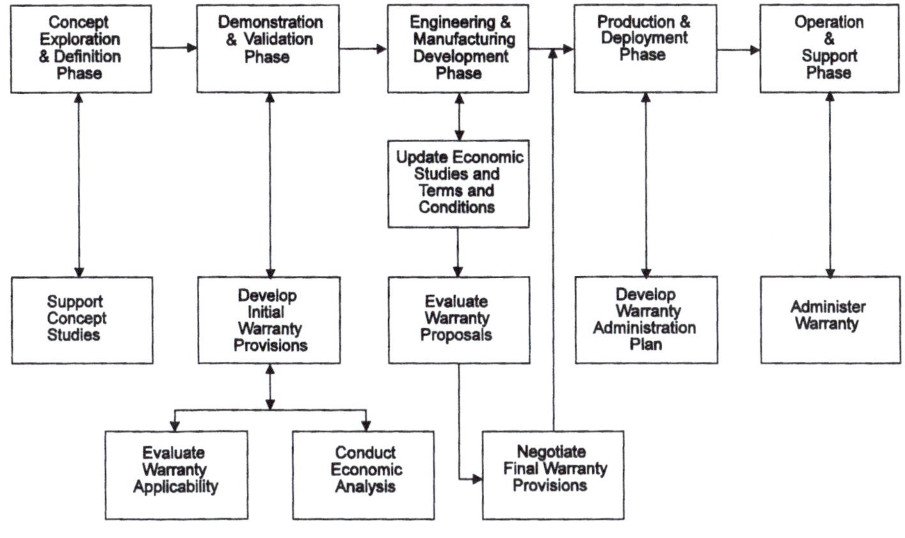

Figure 2.6 Warranty process in the system life cycle.

During the concept exploration-definition phase, the basic design technology options are established. Important to the warranty is the establishment of support concept options to include the number of maintenance levels and the types of support at each level.

In the demonstration-validation phase, the system design concept is established and its feasibility demonstrated. The warranty requirement begins to take shape with the help of inputs from the prospective contractors. Consideration here would be given to which items would require a warranty, possible coverages, feasible lengths, and appropriate remedies for noncompliance. Initial cost-effectiveness analyses are performed and the final warranty applicability determined.

Activity in the engineering–manufacturing development (EMD) phase involves finalizing the warranty requirements suitable for government should-cost analyses and contractor quotes. The contractor's warranty proposal, including cost quote, is evaluated by the government during this phase. Between the EMD and production-deployment phases, the final warranty provisions, including cost, are negotiated between the government and the contractor. In some cases, the contractor's quote is fixed price, and in others it is "not to exceed" (NTE). Experience has shown that the NTE quote at the time of EMD phase is more realistic because of the performance uncertainty of the contractor's equipment yet to be built. In either case the final price is negotiated before or concurrent with initial production.

During the production-deployment phase, the details of the warranty administration are documented in a plan and the systems are deployed in the field. Also, on the basis of the final negotiated price, the cost-effectiveness analysis is completed. Assuming the warranty is cost effective in the operation-support phase, the warranties are put into effect. The contractor administers the warranty, including repair of defective items and monitoring compliance with performance requirements. The government also administers the warranty, including notification of defects to the contractor, processing of defective units back to the contractor, and monitoring compliance to performance requirements.

Not enough can be said regarding the importance of the consideration of warranty issues early in the product life cycle. A win-win warranty likelihood is enhanced considerably when warranties get early and continuous attention.

With regard to consumer and commercial warranties, the warranty process is simpler. Supported by market research and extensive analyses, the warranty statement is written by the seller prior to the marketplace availability of the product. In typical product development cycles, this would take place during either the preproduction or production phases. Once in use, the product would be supported by the seller for the warranty requirements.

Details related to the warranty analysis process are discussed in Chap. 7.

Chapter 3

Warranty Requirements

General Considerations

It is the intent of this section to discuss those factors which influence how a warranty is written. In the consumer sector sellers may use warranties for either promotional or protectional purposes. The warranty strategy may be aimed at improving sales or market share by differentiating the product from its competitors' by some means. Conversely, it may be a defensive strategy to limit the supplier's liability. In the military sector, the contractor responds to written requirements from the customer covering the aspects of a warranty which will ensure adequate coverage for the customer. In some cases, the contractor is encouraged to offer an alternative warranty in addition to responding to the requirement. In these cases the contractor will attempt to highlight factors that seem to potentially have an advantage over the competition and appeal to the customer. An example might be the lengthening of the duration of the warranty over what is required. The contractor may judge the risk to be only minimally increased with the lengthened duration and may be willing to take that risk to offer a more attractive warranty. In offering a more attractive warranty, the chances of securing the contract are enhanced.

Express and Implied Warranties

There are two general categories of warranties:

- Express warranty
- Implied warranty

An *express warranty* is any statement, written or oral, that is made about the product. This includes brochures, advertisements, or salesperson's talks.

An *implied warranty* means what the buyer is entitled to rely on in the absence of a written or oral statement. An implied warranty will provide the buyer with coverage equivalent to what has been done historically in the course of trade. Implied warranties under the Uniform Commercial Code adopted in all 50 states and covering the sale of goods include the concepts of merchantability—that the goods will pass without objection in the trade, and fitness for a particular purpose or use—that the goods will perform the function as claimed.

Warranty Legislation

Consumer and commercial

From a legal point of view, warranties are subject to state and federal laws. The seller's sales management and reliability staff should be aware of the correct legal framework in which to enter their product. Additionally, any formal warranty statement should be reviewed by legal counsel.

The seller must know whether the warranted product is a consumer or a commercial product. Consumer product warranties fall under both federal and state authority:

- The Magnuson-Moss Federal Trade Commission Improvement Act
- The Uniform Commercial Code—enacted by each state

Under the Magnuson-Moss Act, a consumer product is defined to be any tangible personal property which is distributed in commerce and which is normally used for personal, family, or household purposes. Accordingly, automobiles, home appliances, and personal computers are consumer products.

Commercial products are goods sold between merchants. However, if a substantial percentage (greater than 10 percent) of the product sales are in the consumer sector, the product becomes classified as a consumer product and is subject to the Magnuson-Moss Act. Thus, some fairly sophisticated computer printers and private, single-engine airplanes have moved into the consumer sector and become subject to federal legislation.

Full consumer warranties under Magnuson-Moss put many obligations on the seller including, but not limited to the following:

- Remedies for claims
- Service within a reasonable time

- Service at no charge to customer
- Refund of purchase price if defect cannot be remedied

Under a full warranty no restrictions can be placed on implied warranties.

A limited consumer warranty does not meet all the requirements for a full warranty. It is not required that a refund of the purchase price be offered. However, even under a limited warranty, the implied warranty can be restricted only to the same length of time as the written warranty.

If no warranty is stated for a consumer product, the product is assumed to carry a full warranty. This explains why most consumer products have stated limited warranties.

Commercial product warranties under the Uniform Commercial Code may not be as restrictive to the seller. Implied warranties can be totally annulled in a written warranty statement. It is not required that a refund be offered.

The warranty statement is written by the seller prior to the marketplace availability of the product. In typical product development cycles, the statement would be written during either the preproduction or production phases. The statement should be sufficiently detailed so that there is no ambiguity as to the buyer's and seller's rights and responsibilities under the warranty. With many consumer products, warranties are the same for all purchasers. In contrast, commercial warranties may be subject to negotiation between the buyer and the seller. Once agreed on, the warranty requirement becomes part of the purchase agreement.

It is important to remember when writing a warranty, what is not stated is just as important as what is stated. Federal and state regulations must be thoroughly understood. It is therefore strongly recommended that any warranty statement be reviewed by appropriate legal counsel.

Military

Prior to 1984, warranties had been used selectively on various government programs. The reliability improvement warranty (RIW) was used extensively by the Air Force in the 1970s, as was the MTBF guarantee and the LSC guarantee.

The atmosphere prevailing in the early 1980s was one of distrust of government contractors. There had been numerous cases cited of contractor cost overruns on such items as coffee pots, hammers, and even toilet seats, not to mention the military products. Congress was in a mood to steadfastly hold the contractors' feet to the fire to reduce the cost overruns. It should be noted, as an editorial comment, that had the toilet seats not had a 20-ft drop test requirement, the costs might not have been so high. Be that as it may, Congress was in search of a tool to help ensure that the government got its money's worth.

On one of his trips back to his constituency, a congressman from one of the agricultural states happened to be in the market for a tractor. He visited one of the local tractor dealers and was quite impressed with the warranty offered on a new tractor. On his return to Washington D.C., he demanded to know why he "could buy a tractor with a warranty, but DOD can't buy a tank with a warranty."

The wheels were soon put in motion for the preparation of warranty legislation. The 1984 DOD Appropriations Act contained a provision requiring written guarantees for weapon system procurement—Sec. 794. Policy guidelines were issued by then Deputy Secretary Taft on March 14, 1984.

The Office of the Secretary of Defense (OSD) and DOD testified strongly against the legislation as did several industrial associations. Those who so testified cited the fact that DOD had already been requiring warranties on selectively chosen programs and that DOD was in a much better position than Congress to determine the extent of the use of warranties. They further argued that a military tank warranty and a commercial tractor warranty were nowhere near the same. On completion of their testimony, they learned that Congress was in no mood to "let the defense contractors off the hook," especially in light of the above cited widespread publicity.

Effective January 1, 1985, Sec. 2403 of Title 10 of the United States Code revised the law relative to warranties for weapon systems and the appropriate Defense Federal Acquisition Regulations (DFARs) were put in place to implement the law. The 1985 warranty legislation has been a source of great concern to contractors and government customers tasked with making it work in the acquisition process. The government is attempting to understand how to set warranty requirements to satisfy the law. The contractor is struggling with how to quote the warranties in a competitive environment. Both parties are striving to establish negotiating strategies as partners rather than adversaries, and to facilitate warranty implementation.

Title 10, Sec. 2403 of the United States Code is officially entitled "Major Weapon System Contractor Guarantees." For our discussion, *warranty* and *guarantee* will be used interchangeably. Basically, the law requires prime weapon system contractors to provide written guarantees. It further describes the types of coverage required, cites possible remedies, and specifies justifications for a waiver as well as waiver actions. The law suggests that guarantees be tailored to the needs of the procuring activity and the weapon system. A revised DFAR, Subpart 46.770, "Use of Warranties in Weapon System Procurement," was issued as a guidance document for implementing the law.

The law defines weapon systems as "items that can be used directly by the Armed Forces to carry out combat missions and cost more than $100,000 per unit or have a eventual total procurement cost of $10,000,000." The DFAR Supplement (DFARs) provides additional guidance on *weapon system*; it excludes support equipment, training devices, ammunition, and certain commercial items.

The law establishes the prime contractor as having responsibility for the warranty to the government. To reduce their risk, prime contractors may require their subcontractors to provide warranties for the items they supply to the prime item. Under certain circumstances the government deals directly with subcontractors because of the nature of the item. A good example of this is with jet engines. Even though the jet engine contractor is a subcontractor to the prime airframe contractor, the government deals directly with the jet engine vendor on warranties.

Three types of warranties, or coverage, are required by law:

- Design and manufacturing
- Defects in materials and workmanship
- Essential performance requirements (EPRs)

The first two types provide assurance that the product is built to design and manufacturing specifications (structural and engineering plans and manufacturing details) and that it is defect-free in materials and workmanship at the time of acceptance for a specified discovery period. The EPR type guarantees that a product will meet specified performance requirements, usually reliability-related. Maintainability and availability parameters could also be addressed.

EPRs are those measurable, verifiable, trackable, and enforceable weapon system operating capabilities, performance characteristics, and reliability and maintenance attributes that are controllable by the contractor and are determined to be necessary for the system to fulfill its intended mission. Some parameters that are typically covered by an EPR include MTBF, MTBR, operational availability A_0 and repair turnaround time (TAT). The EPR concept is a change from the usual procurement practice by extending the prime contractor's liability beyond acceptance into the operational phase of the weapon system's life cycle.

The EPR guarantee applies only to units in mature full-scale production, units manufactured after the first one-tenth of the total production quantity or after the initial production quantity, whichever is less. It is highly recommended that reliability, availability, and maintenance parameters be chosen as essential since they represent the performance over an extended period of time, unlike parameters such as range, speed, and accuracy which can be verified one time in acceptance testing and would be difficult to verify beyond acceptance. To be effective, EPRs must be carefully selected to include only those requirements which are meant to apply in the operational phase and can be measured and evaluated using existing data that have been collected without imposing an extra burden on maintenance personnel. The warranty should be "transparent" to established maintenance and support procedure.

If MTBF is chosen as an EPR to accommodate the growth in MTBF as is usually experienced during the initial years of deployment, a growth curve may be included in the warranty provisions. The growth curve is recognition that the initial deployment of new technology is usually accompanied by introductory problems.

Provisions may be made for application of penalties and incentives related to the contractor's EPR performance. The penalties and incentives should be significant enough to motivate the contractor to work as hard as possible to meet the EPR. At the same time, penalties should not be so severe as to put contractors out of business if they fail to meet the EPRs. There are numerous examples of where incentives have motivated the contractor to significantly improve the MTBF.

With respect to waivers called out in the law, they can be granted in certain cases when the waiver is in the interest of national defense, or if the warranty, as negotiated, is not cost effective, that is, the cost of the warranty is greater than the life-cycle cost (LCC) savings that would be realized as a result of the warranty. A cost-benefit analysis must be performed to justify a waiver, as applicable.

If a warranted item fails to meet the warranty conditions, the law specifies contractor remedies:

- Contractor must promptly take any necessary corrective action to correct the defect at no additional cost to the government.
- Contractor must pay for reasonable costs incurred by the government in taking the corrective action. The latter is a source of risk to the contractor.

The law allows for tailoring of the warranty to fit the system, procurement, or operational conditions as long as the basic requirements of the law are met. Types of tailoring include the following:

- Negotiation of exclusions, limitations, and time durations
- Use of one-tenth exclusion for second source production
- Extension of coverage and remedies

Risks are implicit in the warranty process for both the government and the contractor. In many cases the risks can be mitigated through appropriate activities during the acquisition phases of the program and through the writing of tailored terms and conditions. In general, the government "bets" that the penalty or incentive features of the warranty will be strong enough to ensure that product performance requirements will be met. The contractor bets that the warranty money paid will remain as profit—the contractor gets to keep the money. Since good quality and performance will win the bets for both parties, this win-win possibility should help structure a warranty where the risks to both parties are acceptable. Some actions to minimize warranty risks are

- Include warranty as part of the acquisition strategy.
- Develop and use criteria to select the correct type of warranty.
- Structure the procurement strategy and the warranty terms and conditions to address the risk factors.
- Perform warranty cost-benefit analyses to determine the "value" of the warranty.

Workable and cost-effective warranties should be as simple as possible. Unduly complex provisions impede interpretation, implementation, administration, and enforcement. In order to prevent confusion from different interpretations of the law, warranty clauses must be clearly written, and leave no room for misunderstanding.

Table 3.1 summarizes the 1985 Warranty Law.

Warranty Structure

Consumer and commercial

There are three basic components in a warranty statement:

- Time period for coverage
- Defects or obligations for which the seller is responsible
- Remedies that will be made if failure occurs within the stated time period

The time period for coverage includes both the length of warranted coverage and the starting point for record keeping. Products which can be easily transported and installed by the buyer might have coverage starting from the date of purchase. Equipment which requires complex installation and/or verification procedures might have coverage beginning with a statement that the equipment was installed in good working order. If items might be purchased as spares or for a facility that is under construction, the buyer would be advised not to accept coverage that began with the purchase, but rather with installation.

TABLE 3.1 Summary of the 1985 Warranty Law

Factor	Definition	Description
Coverage	Weapon systems	Used in combat missions; unit cost is greater than $100,000, or total procurement exceeds $10,000,000
Warrantor	Prime contractor	Party that enters into direct agreement with United States to furnish part of all of a weapon system
Warranties	Design and manufacturing requirements	Item meets structural and engineering plans and particulars
	Defects in materials and workmanship	Item is free from such defects at the time it is delivered to the government
	Essential performance requirements	Operating capabilities or maintenance and reliability characteristics of item are necessary for fulfilling the military requirements
Exclusions	GFP, GFE, GFM	Items provided to the contractor by the government
Waivers	Necessary in the interest of national defense; warranty not cost effective	Assistant Secretary of Defense or Assistant Secretary of the Military Department is the lowest authority for granting waiver; prior notification to House and Senate committees required for major weapon system

The seller usually assumes responsibility for defects in material and workmanship. Any exception must be clearly stated. Common exceptions are appearance items, consumables, any component which has been misused or altered, and equipment which has been serviced by other than the seller or the seller's authorized service personnel.

Common remedies are repair or replacement at no expense to the buyer. The option of a refund exists for consumer products under Magnuson-Moss. Any expenditures for which the buyer is responsible must be clearly stated. Wear items such as tires and batteries may carry a pro rata coverage in which the buyer pays a portion of the cost of the remedy based on usage at the time of failure. The buyer must be informed if used or reconditioned parts are to be employed in the repair.

To avoid confusion between buyer and seller, it is recommended that, as appropriate, the following components be included in the written warranty, in addition to the three basic components cited above:

1. Names and addresses of warrantors—required by Magnuson-Moss
2. Identity of party or parties to whom the warranty is extended:
   - Original purchaser only
   - Any owner within the stated coverage period
   - Direct and indirect customers in a sales chain
3. Countries where the warranty applies:
   - United States and Canada
   - Contiguous United States only
4. The products and parts covered:
   - Computer mainframe only
   - Computer mainframe and peripherals sold as a system
5. Exceptions and exclusions from the warranty terms:
   - Equipment not operated under normal-use conditions as stated by seller
   - Equipment not operated or maintained in accordance with furnished manuals
6. Statement of what the warrantor will do in the event of a defect, malfunction, or failure to conform with such written warranty, at whose expense, and for what periods of time:
   - Remedies that will be taken: repair, replace, and/or refund
   - If used part will be utilized to replace defective part
   - When coverage begins: purchase, installation within a specified time of purchase
7. Statement of what the buyer must do and expenses he or she must bear:
   - Basic maintenance requirements
   - Who pays for transportation, handling, and labor under a warranty claim
8. Step-by-step procedure which buyer should follow in order to obtain performance of any remedy under the warranty, including identification of persons authorized to perform the obligation:
   - Proof of ownership
   - Written description of defect within a reasonable time after the defect became apparent
   - Results of diagnostic tests
   - Steps to be followed in order to ship equipment for repair
   - Name and location of organization responsible for resolving claims

9. The time at which the warrantor will perform any obligations under the warranty:
   - Service center hours for carry-in items
   - On-site days and times for service
   - Charges for off-hours service
10. The period of time within which, after notice of a defect, malfunction, or failure to conform with the warranty, the warrantor will perform any obligations under the warranty. If no time period is stated, "time" is interpreted to mean within a reasonable time so that the product is out of service no longer than necessary for that type of product.
11. Disclaimers, for example, "the warrantor makes no other warranties, express or implied, or of merchantability or fitness for purpose or use of this product."

Military

Recently completed work by the Joint Army-Industry Warranty Working Group (JAIWWG) to improve the value of warranties to the Army revealed a lack of uniformity with regard to the structure of warranty requirements under the law. The JAIWWG made recommendations as to the components that should be included in the warranty statement. The following are the recommended components.

1. Definitions—tailored to the weapon system and warranty type
2. Coverage
   a. Hardware—individual item or systemic
   b. Software—if applicable
   c. Data—if applicable
3. Duration—period of warranty (months, operating hours, miles, etc.)
4. Mandatory requirements (Title 10, Sec. 2403 of U.S. Code)
   a. Design and manufacturing
   b. Material and workmanship
   c. Essential performance requirements (EPRs)
5. Remedies—contractor responsibilities in response to a breach of the warranty requirements
   a. Repair or replacement
   b. Redesign and/or retrofit
   c. EPR penalties
6. Procedure—government and contractor responsibilities relative to the process of warranty implementation
7. Exclusions—specific conditions where the warranty does not apply
   a. Improper handling of the warranted item by the government

   b. Maintenance by the government not in accordance with the approved technical orders
   c. Fire, flood, and other acts of nature
   d. Combat damage
8. Transportation—Army and/or contractor responsibility
9. Other rights and remedies
   a. Government repair of warranted item does not void warranty.
   b. Adjustments due government when the contractor defaults and they accomplish the repair.
   c. Adjustments due government if they determine that repair or replacement of defective items is not required.
   d. The rights and remedies of the government provided for in the warranty do not limit, but are in addition to, the rights of the government under any other clause of the contract.
10. Markings—in accordance with MIL-STD 129/130
11. Warranty cost—a format for the contractor to represent the portion of the contract cost attributable to the warranty:

Contract line item	Administration cost	Repair or replacement cost
1	$	$
2	$	$
3	$	$

12. Warranty data
    a. Status reports—data item description (DID) tailored to exclude unnecessary requirements
    b. Technical bulletins as required

Examples of Representative Warranty Clauses

Consumer and Commercial

> And, its [it's] backed by (XXXXX) standard 3-year warranty on parts and labor.[1]

This statement tells us only that the item has a 3-year warranty on parts and labor. We need to know what "standard" means. This statement lacks sufficient detail for the customer.

> This, and every other (XXXXX) product, comes with an iron-clad 2-year warranty which is backed by a return rate of less than 1% during the warranty period and a very rapid turnaround time.[2]

Here we are told this item has a 2-year warranty, but what is the coverage? We don't know what "iron-clad" warranty means. What does the 1 percent return rate mean to the consumer? Are they guaranteeing TAT? If so, what is the guaranteed turnaround time? This statement lacks sufficient detail for the customer.

> (XXXXX) warrants to the original purchaser that it will repair or replace this product free of charge, if proven to have been defective in original materials or workmanship and returned, delivery costs prepaid, within one year from date of purchase. In no case shall (XXXXX) be liable for any consequential damages for any reason.[3]

The warranty is limited to the original purchaser. The remedies, repair or replace, are stated. The duration is 1 year. The liability is limited. The customer pays for shipment to seller. How is "defective in original materials or workmanship" to be determined?

This statement is a marked improvement over the previous ones, but still lacks sufficient detail for the customer.

> This appliance is warranted against defects in material or workmanship for twelve months from date of purchase. Any problems arising from misuse, dropping or extreme wear are not covered by this warranty.
>
> The store where this item was purchased is authorized to make any exchange only if the return is made within 30 days from purchase date. Returns after 30 days must be made to (name and address). Send the appliance postage paid along with proof of purchase, a note explaining reason for return, and ($XXXXX) to cover handling, insurance, and return postage costs.[4]

The coverage and duration are clearly stated, but what are the remedies? Exclusions are called out, but what is "extreme" wear? The procedure for a warranty claim is clear and thorough.

This warranty statement is clearly the best thus far and provides much more detail for the customer.

> The product you have purchased is warranted by the manufacturer for one year from the date of purchase against defects in workmanship and/or materials. This warranty means that only the parts that prove to be defective during the warranty period will either be repaired or replaced at our option. Should repair become necessary during the warranty period, send your product, postage or freight prepaid, to the nearest Service Center listed on the reverse side of the card. This warranty does not apply if the damage occurs because of (*a*) accident, (*b*) improper handling or operation, (*c*) shipping damage, (*d*) unauthorized repairs made or attempted, or (*e*) use of the product in commercial services. All liability for any indicated or consequential damages for breach of any expressed or implied warranties is disclaimed and excluded herefrom.[5]

With respect to the criteria discussed in the previous section, this warranty statement contains virtually all the necessary factors. This statement contains sufficient detail for the customer, who knows the coverage, length, remedies, exclusions, and procedures for processing the claim. This customer is further apprised that any consequential or incidental damages resulting from a breach of expressed or implied warranties is disclaimed by the seller.

Consumer and commercial warranties have broad-based application. Beyond warranties on appliances, automobiles, cameras, and other standard items, warranties are now offered on education. One community college system offers additional courses, tuition-free, acceptable to the receiving college or university if the community college credits included in an approved degree plan are rejected by the 4-year institution. Additionally, for the career programs, any future employer who finds a community college graduate lacking in job skills and on-the-job competency in their chosen field guaranteed by the community college can initiate a process under which the employee can receive additional training tuition-free. The free tuition is limited to nine credit hours in either case.

Military

> The offerer shall provide detailed build up and rationale of the submitted price. This build up should show the individual build up of all costs attributable to the design and manufacturing portion of the warranty, the material/workmanship portion of the warranty, and each essential performance portion of the warranty.[6]

While this might be useful information to the government, it is virtually impossible for the contractor to execute. This type of detail is not essential for evaluation of the contractor's quote or to subsequently negotiate the warranty price.

> The Subcontractor shall bear the cost of repair/replacement of Depot/General Support repairable parts for warranted failures commencing upon occurrence of the (XXXXX) failure. The Subcontractor's liability shall be limited to a one-time repair or replacement of each failed or defective component, including determination of failure for threshold purpose.[7]

This is an example of a threshold warranty. The government has determined how many failures they will repair and therefore on which failure (i.e., 102d), the contractor's liability begins. Also note that the contractor is liable for only one repair of the warranted item.

> Individual item coverage begins at acceptance of each warranted item and ends 14 months following the acceptance of each warranted item. Systemic defect coverage begins at acceptance of the first warranted item and ends

> 60 months after acceptance of the last warranted item, and includes all systemic defects during this term.[8]

This statement delineates between individual item and systemic coverages, requiring both. Notice the duration of the individual-item coverage is based on each unit's acceptance and lasts for 14 months. On the other hand, the systemic coverage starts at acceptance of the first unit and lasts until 60 months after acceptance of the last warranted item. This is clearly a calendar-based duration. The reasoning is that failures can be linked to an individual unit while defects are linked to root-cause or systemic problems which are applicable across all units to be produced. There must be sufficient time to find and rectify systemic problems, hence the lengthier duration.

> For the warranty period, the warranted items will conform to the essential performance requirements delineated in this contract, referenced specifications, and technical data and amendments thereto.[9]

What we have here is a disaster for the contractor. While this is an easy way to cover the essential performance requirements dictated by the statute, the contractor is left with the impossible task of quoting for the EPR coverage. There is no way to quote coverage on the multitude of EPR alluded to in the preceding requirement. Rather, certain critical-mission parameters, preferably reliability-availability-maintainability (RAM) type, should be called out with guidelines as to how they are to be measured, how often they are to be measured, the criteria for compliance, and the associated penalties or incentives. The cost effectiveness of this warranty is highly unlikely.

> All costs for transportation of items not meeting the requirements of this contract shall be the responsibility of the seller.[10]

Clearly, the seller is responsible for all transportation costs from the field to the repair facility and back again to the field.

> Should the buyer not require repair correction, or redesign, buyer shall be entitled to an equitable reduction in the price of such goods and services.[11]

This is an example of the government's "other rights and remedies." Since the contractor received the up-front warranty money on the basis of a negotiated number of failures, if the government does not require repair of a legitimate failure, the buyer is entitled to a refund.

Warranties that are written per the criteria of the previous section should be adequate to provide the necessary details to help ensure a win-win warranty for the government and the contractor. The following is a complete warranty requirement. This, in the author's opinion, is an excellent example of a clear, concise, and thorough warranty requirement. The reader is encouraged to find an oversight in paragraph 6(c)—no EPRs were listed.[12]

1. Purpose: To delineate the rights and obligations of the contractor and of the government regarding defective supplies delivered by the contractor, and to foster quality performance by the contractor. Specific requirements of this warranty are presented below.
2. Scope: This warranty is directed toward the correction of all defects that exist in the warranted items at the time of acceptance or that are injected into the warranted items by the contractor during the repair or replacement of the items. This warranty is also directed toward correction of the failure of the warranted items to meet their essential performance requirements.
3. Definitions:
   (a) Acceptance: The execution of an official document (DD Form 250) by an authorized representative of the Government.
   (b) Defect: A condition or characteristic of the warranted items that is not in compliance with the requirements of this contract. A defect does not necessarily affect performance.
   (c) Failure: Breakage or malfunction of a part, or damage to a part, which renders the warranted item unserviceable, or a condition which causes or would cause a warranty item to fail to meet its performance requirements. A failure may also be a defect.
   (d) Failure-Free Warranty: A warranty that provides a period of time during which the contractor either repairs or replaces all warranted items that fail or are defective at no additional cost to the government.
   (e) Handoff: The issuing of the end item to the field (user) as verified by the applicable handoff document.
   (f) Repair: The elimination of a defect.
   (g) Systemic Defect: Defects of a repetitive nature which are of the same root cause and affect a specific group or population of items.
   (h) Warranted Items: The system and each component thereof.
   (i) Warranty Administrative Cost: The cost incurred by the contractor in administering the provisions of this warranty. It does not include the labor and parts cost nor the associated overhead cost for repairing or replacing warranted items.
4. Warranty Coverage: This warranty provides both individual item coverage and systemic defect coverage as follows:
   (a) Individual Item Coverage: Warranty coverage that requires individual warranty claim actions for each failure and defect. These claim actions will only be made on a warranted item when the item or a component of that item is to be repaired or replaced at the depot level.
   (b) Systemic Defect Coverage: Warranty coverage that provides protection to the lowest level of impact or expense and requires a contract remedy in accordance with this contract. This coverage applies to all warranted items.

5. Warranty Periods: This warranty provides two distinct warranty periods as follows:
   (a) Individual Item Coverage: The period begins with the acceptance of each warranted item and ends 36 months following acceptance, or 24 months after handoff, whichever occurs first.
   (b) Systemic Defect Coverage: The period begins with the acceptance of the first warranted item and ends 36 months after acceptance of the last warranted item.
6. Warranty Requirements. Notwithstanding inspection and acceptance by the government of the warranted items furnished under this contract or any provision thereof, the contractor warrants for the warranty period that the warranted items:
   (a) Will conform to the design and manufacturing requirements of this contract and any amendments thereto.
   (b) Are free from all defects in material and workmanship when accepted by the Government.
   (c) Will conform to the following essential performance requirements.
7. Warranty Procedure.
   (a) The contractor shall provide the materials and services necessary to implement this warranty and shall assign a single point of contact to communicate with the Government on all warranty related issues.
   (b) Individual Item Coverage: The Contracting Officer will provide written notification to the contractor within 90 days of the date of discovery of any warranted item that fails or is defective. The contractor, notwithstanding any disagreement with the Government regarding this warranty, shall promptly comply with the direction provided by the Contracting Officer regarding the defective items at no increase in the contract price. Unless agreement has been obtained from the Government to do otherwise, repair or replacement shall be accomplished within 60 days after receipt of the defective items by the contractor.
   (c) Systemic Defect Coverage: In the event a warranted item does not meet the requirements of this contract and the defect is systemic, the Contracting Officer will provide written notification to the contractor within 60 days of the date the defect is determined to be systemic. Coverage is invoked when notification of the systemic defect is received by the contractor. The contractor shall then prepare a corrective action plan in accordance with DI-RELI-80254. Upon Government approval of the plan, the contractor shall take the designated corrective action and is liable for the cost thereof, including the cost of preparing the plan. The corrective action shall provide for the repair, replacement, or retrofit of the warranted items on an inventory-wide or total asset basis when applicable. This remedy shall include redesign of the item if redesign is necessary in order to meet the requirements of this contract.
   (d) If it is later determined that the defect was not subject to the provisions of this warranty, the contract price will be equitably adjusted.

(e) The Government or contractor shall prepare Warranty Status Reports in accordance with DI-MISC-80733.

8. Warranty Status Reports: The contractor shall prepare Warranty Status Reports in accordance with DI-MISC-80733.
9. Transportation Cost: When warranted items are returned to the contractor in pursuance to this warranty, the contractor shall bear the cost of transporting the items to the contractor's repair site and back to the user.
10. Exclusions:
    - (a) The contractor shall not be obligated to repair or replace warranted items if the facilities, tooling, drawings, or other equipment necessary to accomplish the repair or replacement are not available to the contractor due to action by the Government. In this case, if repair or replacement is directed by the Government, the contractor shall promptly provide written notification to the Contracting Officer of the nonavailability.
    - (b) For the purpose of this warranty, the term "Performance Requirements" does not include performance characteristics that are described as goals or objectives.
    - (c) The provisions of this warranty do not apply to:
        1. Items damaged in combat.
        2. Liability for loss, damage, or injury to third parties.
        3. Items with defects which are beyond the control of, and not attributable to, the contractor. Such items are those which have been damaged through willful misconduct, accident, misuse; abuse; improper installation or application; unauthorized maintenance or maintenance not executed in accordance with applicable technical manuals; negligence in transportation, handling, or storage; catastrophic damages such as fire, flood, or explosion; or an act of God.
        4. Items with faded or chipped paint, scratches, dents, nicks, or any other cosmetic damage resulting from normal and customary use.
    - (d) Government-furnished property (GFP) shall be warranted only to the extent of proper installation so as not to degrade the warranted item's performance or reliability. If the contractor modifies or otherwise performs work on the GFP, this warranty shall cover such modification or other work.
11. Other Rights and Remedies
    - (a) The Government shall not be responsible for any extension or delay in the scheduled deliveries or periods of performance under this contract as a result of the contractor's obligation to repair or replace defective items. Furthermore, there shall not be any adjustment of the delivery schedule or periods of performance as a result of the repair or replacement of defective items unless provided for by the inclusion of a supplemental agreement, with adequate consideration to the Government, to this contract.
    - (b) The rights and remedies of the Government provided for in this warranty do not limit, but are in addition to, the rights the Gov-

ernment has under any other clause of this contract. Disputes arising regarding this warranty will be resolved in accordance with the "Disputes" clause of this contract.

(c) The rights and remedies of the Government provided for in this warranty shall survive final payment.

(d) This warranty shall not be voided by any Government performed repair of any warranted item when accomplished in accordance with standard military service maintenance procedures. This includes the substitution of parts procured by the Government from any other source. However, the contractor shall not be responsible for the substituted part or any resulting damage caused by the substituted part.

(e) If the contractor fails to repair or replace defective items in the time specified in this warranty and unless an extension is granted by the Contracting Officer, the contractor shall pay the costs incurred by the Government in procuring the necessary parts and in accomplishing the repair or replacement.

(f) If the Government determines that repair or replacement of defective items is not required, the Government shall be entitled to an equitable adjustment in the contract price.

12. Markings: Each warranted item delivered under this contract that is depot repairable/recoverable shall be stamped or marked in accordance with MIL-STD-129 and MIL-STD-130 and as follows:

(a) A warranty identification label or plate shall be securely applied or fastened to each warranted item in a conspicuous location.

(b) Background marking shall be of alternating blue and neutral (natural color of material) 45 degree diagonal stripes of equal width. The width of each stripe shall be approximately equal to the character height. The blue color shall approximate FED-STD-595, color number 35250. The plate or label shall contain the following data:

Warranty Item
(Title in Bold Letters)

NSN:____________________
(National Stock Number)

EXP:____________________
(The warranty expiration date shall be expressed as month, day, and year)

CAGE:____________________
(Commercial and Government Entity Code, i.e., 96906)

Contract:____________________
(Contract Number)

WTB:____________________
(Warranty Technical Bulletin Number)

13. Warranty Cost:
   (a) The amount indicated below represents the portion of the contract cost attributable to this warranty:

Contract Line Items	Administrative Cost	Repair/Replacement Cost
______	$ ______	$ ______
______	$ ______	$ ______

   (b) If the warranted items delivered under this contract are ever designated for long-term storage (such as war reserves or preposition stocks) or for foreign military sales, the contract price will be equitably adjusted.

Contractor Strategy

Relative to the interpretation of the warranty law, warranty statements designed to satisfy the law cover a wide range of requirements. Contractors are faced with understanding, quoting, and administering a plethora of warranty requirements. Since the law is subject to many interpretations, contractors must establish a systematic strategy and build an experience base to be responsive. Figure 3.1 presents the contractor strategy equation.

The contractor strategy equation is a function of four important actions which, when summed or accumulated, equal or yield a warranty experience base necessary to be responsive to warranty requirements. This strategy allows the contractor to benefit from their warranty experience on each successive warranty effort. The goal of this strategy is to be better prepared and more responsive to the existing set of warranty requirements than for the previous set of requirements.

At Texas Instruments, through the joint efforts of the Quality and Contracts organizations, a *Manager's Guide to Warranty Legislation* has been prepared and updated.[13] The *Guide* presents a summary of the law and its exclusions, basic statistical considerations relative to reliability, a model warranty for use in cases where alternative warran-

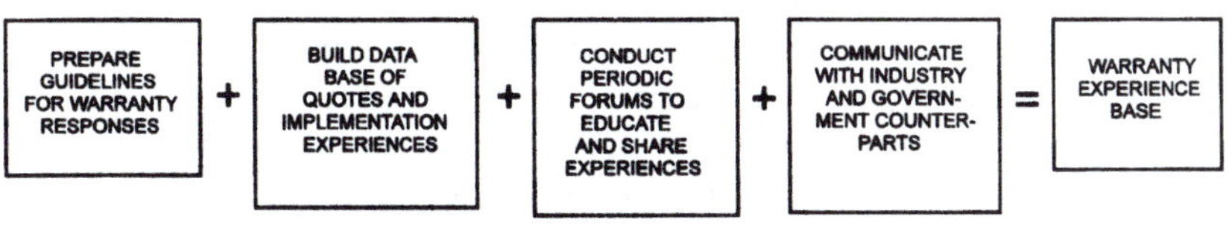

Figure 3.1 Warranty strategy for the contractor.

ties are proposed, a history of warranties at TI (including the customer and the program manager), and a manager's warranty checklist consisting of some 25 points for consideration. The *Guide* was distributed to some 500 managers within TI and has received plaudits from several of the users as to its value as a roadmap.

It would be most desirable to have historical data on what each warranty actually cost the contractor. This data is difficult to secure and compile. In lieu of that, a database of warranty quotes tied to specific requirements is also helpful. Along with the quote value, the method of estimation is also included. With this information available, a starting point can be established for the existing warranty rather than starting from scratch. If this data can also be tied to a particular customer, additional benefits relative to consistency can be established and maintained.

With the rapidly changing workforce at TI we found it necessary to conduct periodic seminars at the various sites to keep our managers and practitioners current on the warranty law and provide insight on how to be responsive to warranty requirements. These seminars were limited to half-day or less in length since it is quite difficult to capture managers for any longer period. The response is excellent, with some 300 attendees total from the three major sites. Not only have the seminars been instructional in content, but several attendees were surprised that a warranty capability existed within the company and were eager to employ it, some immediately.

Through attendance at seminars and participation on committees, we have been able to identify and communicate with our industry and government counterparts. This liaison has been invaluable in terms of responding to particular warranty requirements from a specific government office. We have been able to discuss the requirements and the associated difficulties in responding with the customer's warranty agent and not "guess" at the meaning of the requirements. Being able to compare warranty experience with our industry counterparts has paid dividends relative to new and effective analysis techniques. A "closed shop" with regard to warranties and their attendant subjectiveness is not a wise strategy. Contractors need all the help they can get in order to be responsive.

The accomplishment of these four activities should produce a solid warranty experience base for the contractor. Armed with this experience, the contractor's ability to be responsive should be significantly enhanced. None of these four activities is particularly time-consuming or expensive, but collectively they provide a strategy that is sufficiently sound to better prepare the contractor for the next warranty requirement.

Importance of Warranty Requirements

Not enough can be said regarding the necessity for clear, concise, and simple warranty statements. The clarity of the warranty statement is crucial if the warranty is to be a win-win situation for both the customer and the supplier. Effort expended early in the development cycle to clearly define the warranty requirements for a particular product can pay rich dividends when the warranty is implemented.

References

1. Tektronix, scope and sales brochure, 1989.
2. Leader Instruments, scope and sales brochure, 1989.
3. Pittway Corp., sales brochure, 1990.
4. Vidal Sassoon, sales brochure, 1990.
5. Sears Co., sales brochure, 1990.
6. U.S Air Force, request for proposal, gunship aircraft, 1986.
7. U.S. Army, request for proposal, fire control system, 1987.
8. U.S. Army, request for proposal, fire control system, 1988.
9. U.S. Air Force, request for proposal, avionics system, 1988.
10. U.S. Air Force, request for proposal, avionics system, 1988.
11. U.S. Army, request for proposal, fire control system, 1989.
12. U.S. Army, request for proposal, fire control system, 1990.

Chapter

4

Fixed-Price Extended-Duration Repair Warranty: Reliability Improvement Warranty

General

In this chapter, the reliability improvement warranty (RIW) is discussed as an example of a fixed-price extended-duration repair warranty. RIW has been used extensively by the government since the mid-1970s, particularly for electronic systems. The objective of the RIW is to achieve acceptable reliability while providing the motivation and mechanism for reliability improvement. The RIW, by our previous definition in Chap. 2, is classified as an incentive warranty. The incentive is accomplished through a fixed-price contract provision for the contractor to perform repair for all covered failures during the warranty period. The negotiated price for the warranty is based on reasonable costs to repair covered failures and a field MTBF consistent with that expected. Using our previous Chap. 2 example, if the warranty is for 200,000 operating hours and the expected field MTBF is 1000 h, assuming the contractor provides equipment that meets the expected MTBF, the quantity of failures expected to occur is 200 (200,000 ÷ 1000). The 200 failures becomes the basis for negotiating a warranty price.

The contractor, who has already been paid up front for the warranty, is interested in developing and producing equipment with an MTBF greater than 1000 h if the additional development and production costs to accomplish this reliability improvement are less than the reduction in future warranty repair costs. The contractor, who is responsible for repairing all failures, has the opportunity to devote resources to detect and fix systemic failures as early as possible. If this can be done in time to reduce the number of future failures and this savings is more than the invested cost, the contractor will be motivated to do it.

As stated previously, RIW has been used by the government in the past on electronic systems. The RIW approach has required changes to support systems, but has proved to be administratively workable. The Air Force, particularly on the F-16 aircraft avionics, has used RIW more extensively than the other services.

Definition and Scope

- An RIW is a fixed-price commitment for the contractor to perform repair services for an extended period of time with the objectives of improving reliability and reducing support costs.
- An RIW provides a contractual incentive for the contractor to improve reliability through early design emphasis and no-cost engineering change proposals (ECPs) to reduce the repair and support costs. The contractor's profits are tied to field reliability.
- An RIW should contain a reasonable balance between contractor risks and incentives.
- The greatest value of an RIW is realized in the initial years of field deployment.

According to Chap. 3, under a basic failure-free warranty, the contractor must correct all failures that occur during the warranty period. This extended period of time is usually greater than 2 years and could be as long as 5 years. For that period of time the contractor is the depot and is responsible for the maintenance and upkeep of the systems produced. The RIW concept is that if contractors are responsible for care of their own equipment, they will produce or upgrade their products with a higher level of reliability and customers will therefore realize the benefits in terms of improved reliability and reduced support costs.

It is assumed that the contractor, who essentially gets paid up front in a fixed-price contract, will want to keep as much of that money as possible. If early on in the warranty the rate of returns of faulty units is higher than the expected number negotiated with the customer, it may be more cost effective for the contractor to effect an ECP than to continue to repair the excessive number of failures for the remainder of the warranty period. This is a judgment the contractor must continually make during the duration of the warranty. If a major portion of the warranty period remains, the decision is simpler. The closer to the end of the warranty period, the tougher the call. The contractor's profits are clearly tied to the field reliability of the product.

As discussed in Chap. 2, warranties involve risk for both supplier and customer. In the RIW the supplier can be asked to assume a high level of risk since the product reliability estimates are made without the benefit of testing and, in most cases, before the product is built. Of course, customers always take risks on the products they buy. The key is a balance of risks and incentives for the contractor through effective negotiations. In other words, is it worth going through the engineering and production effort for an ECP when perhaps the reliability will grow anyway on the basis of the feedback of information on failures and the incorporation of this information into the units yet to be fielded under the warranty.

Since most of the problems associated with the design occur early in the field use period, the greatest value of an RIW is realized in the initial years of field deployment, specifically during the warranty period. Additionally, if the design changes are made early enough in the warranty period, the majority of the units yet to be produced can benefit from the ECPs.

Potential Benefits

Government benefits

- Responsibility for field reliability and repair rests with the contractor
- Life-cycle cost (LCC) is controlled
- Minimal initial support investment is required

In procurements without a warranty the contractor's responsibility ends when the equipment is accepted by the government. While contractors may be motivated through customer satisfaction and basically taking pride in what they develop and produce, there is not the additional motivation of being held accountable for the field reliability and repair of their equipment through the early years of deployment.

A popular question arises often. Why should the government even need a warranty when all they are asking is for the contractor to meet the contractual requirements? The contract should be motivation enough for the contractor to develop and produce good equipment. While all contractors would say that they fully intend to meet the contractual requirements, the knowledge during development and production that they will warrant their equipment and indeed be held accountable for the field reliability and repair provides the additional monetary teeth to perform to their best potential. Carrying this discussion a bit farther, why should the government pay for a warranty? Keeping in mind that a contract without a warranty does not require contractors to repair their own equipment, there is attendant cost with the warranty for repair and administration for which contractors should be paid. The bottom line is that warranties are necessary to fully motivate the contractor and the contractor should be paid for at least the basic costs of warranty administration and repairs.

Up to this point we have not discussed the incentive aspect of the RIW. Warranties do provide motivation for contractors, but providing an incentive within the warranty for them to do better than what they negotiated can pay rich dividends for the government in terms of improved performance and reduced support costs. Contractors are motivated to perform the necessary trade studies in development and production leading to higher reliability equipment in the field initially and when the government assumes depot repair of an item at the conclusion of the warranty. The possible repair cost savings to the contractor is the catalyst.

In Chap. 2 we mentioned the cost-effectiveness analysis required of the government prior to warranty implementation. The government evaluates the LCC with and without a warranty. If the warranty is not cost effective, the government can seek a waiver of the warranty through DOD. If the warranty is cost effective for the government, the negotiated warranty is implemented. LCC is a very important consideration for the government. There is potential for a significant reduction in LCC. LCC experience shows that for most systems, the operating and support costs drive the LCC. Going a bit farther, reliability is generally the key contractor controllable parameter. If the contractor manages reliability such that the costs associated with reliability are traded off throughout the development and production phases, the acquisition costs and support costs can be balanced such that the total LCC can be controlled.

With the tightness of the defense budgets, it is attractive to the government that the contractor be the depot for the initial years of deployment. Outfitting a new depot or even modifying an existing facility is quite costly. The test and support equipment and the necessary spares, not to mention other auxiliary equipment, require a significant initial investment. In many cases, the government, aside from monetary considerations, would not be physically able to bring a depot on line. There is a tendency currently within the services to use the contractor as the depot; it is generally called *interim contractor support* (ICS). The warranty is an extension of ICS to embody the fixed-price approach and the requirements of the law. The RIW further extends the warranty to incorporate the incentive aspect. The RIW fits nicely within the monetary and physical constraints of the government.

Contractor benefits

- Increased profit potential realized if MTBF can be improved above pricing base
- Multiyear guaranteed repair business which is prepaid
- Familiarity with operational reliability and maintainability aspects of their equipment
- Improved reliability of contracted equipment

The ultimate aim of the contractor is to maximize profits. Within the RIW, the contractor's goal is to keep as much of the up-front warranty money as possible. By improving the field MTBF above that which was negotiated, the contractor gets to keep more of this money because of the reduced number of repair actions. Here again, the assumption is that the savings in reduced repair costs is more than the "extra" development and production money spent to improve the reliability. Engineering change proposals (ECPs) can be costly, and the contractor must continually evaluate the costs against the potential reliability improvement and the associated repair and support cost savings. This is a challenging exercise and involves risk to the contractor if the ECP investment does not, in fact, improve the field reliability any appreciable amount. The incentive to produce higher reliability is definitely present for the contractor with the RIW, but the contractor needs to be realistic relative to the costs and the expected gain of any ECP that is implemented.

With the RIW, the contractor is assured of 3 to 5 years of prepaid repair business. Without the RIW, either the contractor conducts a repair contract with the government and is paid on a piece-rate basis for the repair work, or the government has someone else do the repairs. With the obvious push toward other than government repair depots, the RIW offers the contractor the benefit of getting paid for these repairs up front in a fixed-price contract. The RIW also guarantees the repair business for the contractor. This guarantee allows the contractor the visibility for staffing and the attendant training, and equipping the depot with the knowledge that this effort will go on for, say, 3 to 5 years. If the contractor has costed the repairs adequately in the RIW quote and it was negotiated reasonably, this could be a significant boost to the overall profit.

Not enough can be said regarding the importance of firsthand knowledge of the reliability and maintainability aspects of one's own equipment. With this extended period under the RIW the contractor has the opportunity to find reliability and maintainability problems and implement the necessary corrective actions to improve the reliability and maintainability. Being the depot provides opportunities to gather data and accurately assess the details of the problems, which makes the tradeoff analyses relative to ECP execution more meaningful, with a higher likelihood of a cost-effective choice. This improved visibility of reliability and maintainability problems can pay great returns when considering follow-on business, particularly with that customer. Armed with this heightened awareness, the contractor can enhance future sales by using this acquired knowledge not only on the current program but future programs as well.

Customer satisfaction, contractors are learning, is a very important factor in their existence. Reliability happens to loom quite large as a factor in customer satisfaction. If a contractor can improve equipment reliability over what was required or negotiated, the positive impact is far-reaching. For example, Texas Instruments builds the High Speed Anti-Radiation Missile (HARM) for the Navy and Air Force. HARM was the first major weapon system acquired under the warranty law. The warranty was bid and negotiated in good faith. The field MTBF far exceeds that which was negotiated. Additionally, it performed excellently in Desert Storm neutralizing the air-defense radars. The contracts and logistics people who negotiated the warranty are very unhappy because they could have negotiated a higher MTBF. The operational people, on the other hand, are delighted at the HARM's overall performance and could not wait to inform Congress of the success. The achieved reliability of HARM far outweighed "lost" negotiations as far as the Navy and Air Force were concerned. It is the author's opinion that the success of HARM has, and will continue to have, a profound impact on future procurements for Texas Instruments, not only with regard to missiles but for other equipment and applications as well. Contractors would do well to seriously consider investments in reliability. As pointed out in Chap. 12, customer satisfaction will become even more important as we move into the twenty-first century.

Potential Risks

Government risks

- RIW price
- Reduced self-sufficiency
- Administrative complexity
- Transition from contractor to government maintenance

There is a possibility that the government may pay too much for the RIW coverage. If the RIW is not negotiated thoroughly, the government may not realize maximum benefit from their investment. It is important that the government clearly understand the costs in the contractor's quote. If the contractor has included money for ECPs, the government must clearly understand what they are getting for the ECP expenditure. If the contractor has included risk money relative to the MTBF, the government must know how the risk dollars were determined. If these considerations, and others, are clearly uncovered and understood, the likelihood of the government paying too much for the RIW is greatly reduced.

Because the government is tied in with one contractor in a warranty, their self-sufficiency is at risk. Not only are they tied in with one contractor; they are locked in for 3 to 5 years. The contractor's performance can be negatively impacted by such things as business losses, strikes, and other unplanned occurrences. It would be quite difficult for the government to "change horses" to another contractor well into the warranty if the contractor cannot execute the warranty for the preceding, or other, reasons. Without a viable alternative, the government is forced to work with the afflicted contractor to implement the warranty. The chances for a successful warranty are severely hampered.

RIWs are complex to administer. The ECPs that the contractor proposes must be approved by the government before they can be implemented. The government must be sure that the ECP is in their long-term interest and therefore requires time and effort to make that determination. Being the depot, the contractor is required to submit status reports on repairs and corrective actions. These must be monitored by the government and any deficiencies reported and followed up on with the contractor. If the added complexity and its associated cost are not fully comprehended by the government, there could be a greater expenditure than was anticipated and negotiated, not to mention a possible shortage in personnel to adequately administer the warranty.

We noted earlier that a possible benefit to the government associated with RIWs was reduced initial support investment. Along with this, however, is the possibility that the transition from contractor to government (organic) depot could be more costly than if the transition were not necessary, that is, organic depot from the start of the warranty. Often this tradeoff is moot because the government is physically not able to assume the depot maintenance role. The government must, however, be made aware of this potential cost risk.

Contractor risks

- Possibility of losing a significant amount of money
- Commitment to a fixed-price agreement with limited reliability data
- Pricing warranty low to keep LCC low to get the contract—possibility of losing money on RIW option
- Pricing RIW high to make a profit on RIW option—possibility of losing contract because of high overall LCC

We have thoroughly discussed the incentive aspects for the contractor. We need now to look at the other side of the coin. Suppose the achieved field reliability is below the negotiated value. In this scenario, the contractor will be repairing more, perhaps many more, failed items than anticipated and may have negotiated a repair cost that could be significantly lower than what is actually expended. The bottom-line result is that the contractor's own money is used to effect the repair because the money received up front was not enough. A possible compounding of the problem occurs if the contractor also executed several ECPs, not quoted, to improve the reliability and got limited or no field reliability improvement. This is a very real risk to contractors, so much so that some are reluctant to enter into such as an RIW agreement, viewing the risks as dominant over the incentives.

This next contractor risk is the one that gets the most attention. In order to fully comprehend the magnitude of this risk, we must realize that contractors may be asked to enter into fixed-price agreements as early as the engineering–manufacturing development (EMD) phase of a program. At that particular point in time, they have a system design and have demonstrated the concept with prototypes. They have not yet built this equipment and have limited or no reliability data. They must base their reliability commitments on allocated and possibly predicted reliability. Having limited knowledge of how their production units will perform in the field, except perchance for analogy to a previous system, they face a real dilemma. The only real hedge they have is through pricing risk into their quote. Rarely do they feel comfortable with meeting the required field MTBF and, therefore, pricing some risk if they don't meet the requirements is a means of covering themselves. They are not saying they will not meet the required MTBF; they are saying that having not built the equipment, they are at risk to meet the requirements.

The next two risks are related. In most cases, the award of a development contract is based, in some way, on the expected LCC of the contractor's proposed design. A very significant part of that LCC estimate is the cost of the warranty. There are two options here. The contractor can (1) purposely price the fixed-price warranty low to keep the LCC low to increase the chances of winning the contract and risk the possibility of losing money on the RIW option, or (2) price the RIW high, thus ensuring a profit on the RIW, and risk the possibility of losing the contract because of high overall LCC. It can be argued that the first option is preferred since it is more important to get the contract than worry about making a profit on the warranty. If we were able to interview some contractors, they might take exception to this since they lost a significant amount of money on the RIW. Ideally, a contractor may not have the lowest LCC and therefore have not totally sacrificed themselves on the RIW, but have other redeeming features to their proposed designs and win the contract on the overall strength of all factors. This dilemma has caused many a contractor great grief when bidding an RIW in conjunction with LCC. At best, this is a judgment call.

Potential Problem Areas

- Changing roles of government and contractors from adversaries to partners
- Reluctance of contractor program managers to emphasize the profit potential in RIWs
- Conducting the necessary tradeoff analyses to provide adequate pricing visibility
- Providing an adequate database for RIW pricing

The win-win possibility with warranties is made or broken at the negotiating table. If both contractor and the government are not totally open with each other, the likelihood that both will benefit is small. Traditionally, contractors have been reluctant to share all their data or assumptions with the government. It is imperative that all the assumptions and data be on the table so that both sides can determine their true positions. There is no magic formula for this required partnership. The best solution lies in the education of both sides that it is possible for both to be winners if and only if there is an openness of communication. There have been enough successful warranties to indicate that this partnership relationship is rewarding.

My experience is that program managers will invariably choose the dark side of warranties and dwell on them rather than consider the profit potential with the RIWs. Here again, education is really the best way to overcome this tendency. Once they understand that increased profits are possible with the RIW, they may change their perspective. It was not until recently that companies even viewed warranties as possible sources of profit. Since warranties are part of their business, why shouldn't they be viewed as opportunities for profit? "Just cover yourself" was the war cry of contractors relative to government warranties for many years. I have observed an awakening of management and an aggressiveness toward warranties most recently. I believe the emphasis on customer satisfaction related to the total quality thrust has helped change some mind-sets relative to warranties.

We have discussed the necessity for extensive tradeoff analyses on the part of the contractor relative to the cost effectiveness of ECPs to improve the field reliability of warranted units to reduce future repair costs. Many times these tradeoffs are not adequately performed to provide sufficient visibility for knowledgeable decisions. In the long run, the result is seldom beneficial to the contractor, who may have either underestimated the cost of the ECP or overestimated the reliability improvement on incorporation of the ECP. Tradeoff analyses are the cornerstone of a development program, and they must be performed as thoroughly as possible, utilizing the best data available. With the high dollar value riding on warranties, they demand a concerted effort to provide a decision tool for management.

The last item is a sister to the previous item. There never seems to be sufficient data to support decisions. Warranty pricing is no different. The timing of the fixed-price commitment attendant with RIW will rarely ever provide the opportunity for sufficient data. The reader is referred to Chap. 3, on contractor strategy. While contractors may not have sufficient data for the equipment under consideration, if they have established and maintained files of the history of quotes related to warranty requirements from particular customers, they may find some useful data to support their current quote. Insufficient data will be a cry as long as contractors build equipment. It cannot, however, be an excuse for the contractor. As discussed earlier, there are ways for the contractor to cover the uncertainty arising from limited data during the performance of the RIW quote.

Application Criteria

- Equipment having potential for reliability improvement and support cost reduction
- Competitive procurement
- Compatibility with fixed-price contract award
- Equipment in production over a substantial portion of warranty period
- Large quantity procured and high expected operating hours
- Moderate to high expected initial support costs

Since the core of the RIW is the contractor's incentive to improve reliability to reduce repair costs, the equipment in question must be such that these aims are possible. A mature system design with minor modifications to satisfy this particular program would likely not be a good candidate since a major portion of the reliability growth has already been achieved. On the other hand, a new technology to accomplish the equipment's mission would likely be a good candidate.

The need for a competitive procurement is quite obvious. The competitive procurements typically foster more realistic contractor pricing.

The more realistic the contractor pricing, the higher the likelihood of a cost-effective warranty for the government. Without competition, the LCC and RIW pricing tradeoff discussed earlier has no teeth, and the lone contractor will likely produce a higher RIW quote.

It is important that the RIW be tied to a fixed-price contract. Contractors getting up front all the money that they are going to get is a key factor in making the RIW, with its attendant incentives, effective. A cost-plus contract would not be as effective in accomplishing the government's goals through RIW as would the fixed-price contract.

This next criterion is crucial to the success of the RIW. There must be the opportunity for the contractor to implement the ECPs necessary to achieve the improved field reliability in subsequent production units. This forward-fit concept is much more economically desirable than back-fit of already fielded units. In many cases, due to typical production schedules, this criterion is not difficult to satisfy.

The large quantity of systems procured and the high expected operating hours work in concert to produce a large number of failures requiring repair during the warranty period. This high expected repair cost fits in nicely with the first criterion of significant support cost reduction. Without a large quantity of repairs, the contractor may opt to continue with the existing level of reliability since the payoff in reduced repair costs is not significant enough to offset the ECP investment(s).

The final criterion is a modifier of the previous one. The benefit of RIW to the government is maximized when the support costs are expected to be fairly heavy from the initial fielding of the systems and throughout the warranty period.

RIW Case Study

The following is an actual case study on an Air Force–procured TACAN (tactical air navigation) avionics system. The system was a solid state design, had built-in test (BIT) capability, and, despite a challenging reliability requirement, had potential for significant reliability improvement. The following are key considerations to the application and conduct of the RIW:

- Three LRUs, 2600 parts
- 4-year RIW with a MTBF guarantee rider
- Meeting the key application criteria:
  - Competitive procurement
  - Sufficient procurement quantity and production extended over most of the warranty period
  - Potential for reliability improvement and support cost reduction

Chapter

5

Performance Guarantees

Definition

Performance guarantees are contractual instruments used to ensure that desired levels of the essential performance parameters (EPRs) are met. These levels are demonstrated on full-rate production units under field conditions. The demonstration of the EPR levels can take place in a controlled demonstration test or in a normal operating scenario. These performance guarantees can cover operational parameters such as speed, range, accuracy, and thrust, and/or reliability, availability, maintainability (RAM), and cost parameters. Ideally, these performance guarantees should be written as to provide incentives as well as penalties for the contractor. Performance guarantees should be designed to supplement the fixed-price extended-duration repair warranties such as the RIW in order to provide thorough warranty coverage under the law. As briefly discussed in Chap. 2, it is imperative that the warranty requirement specify clearly which parameters are designated as essential and hence require measurement and validation, how and when the parameters are to be measured, the criteria for compliance or noncompliance, and the method for determining the incentives or penalties associated with compliance or noncompliance.

Types

The operational parameters will be discussed first. The more readily specified RAM parameters will then be discussed in detail relative to four criteria: objective, description, applicability, and measurement.

Operational parameters

These are parameters generally associated with the adequacy of the system design to meet mission performance requirements. They include such parameters as speed, range, resolution, accuracy, thrust, and fuel consumption. Typically these capabilities are demonstrated in acceptance testing and are basically not time- or use-dependent. It is the author's opinion that these parameters should not need to be guaranteed in the postacceptance field use period. Customers get very little added coverage when these are included and may pay too much for what coverage they do get. The contractor is subjected to greater risk with these operational EPRs. For example, suppose Army PFC Roger Doright is working in the motor pool. He has learned through discussions with the maintenance officer that the new Haul-More truck is guaranteed to stop within 60 ft when traveling at 45 mi/h. One day in the discharging of his normal duties he has occasion to drive the Haul-More truck. He remembers the guarantee and promptly puts the truck to the test. He carefully adjusts the truck speed to 45 mi/h and applies the brakes. He jumps out, measures the distance from where he thinks he applied the brakes to where the truck actually stopped, and, lo and behold, the distance measured is 62 ft. Excited as he can be, he reports the deficiency to the maintenance officer, who completes the deficiency report and processes it up the chain of command. When the dust settles, the contractor has been ordered by the government contracting officer to perform a costly redesign and retrofit effort to correct the defect and demonstrate that the truck can, in fact, stop within 60 ft when traveling 45 mi/h.

This example may seem a bit far-fetched, but clearly makes the point that with operation-capability EPRs, the contractor's risk increases drastically, with the increased risk go dollars to cover the contractor in the quote, and finally, the cost effectiveness of the warranty for the government is jeopardized.

The bottom line is that one-shot mission performance capabilities are difficult to control in the field, are adequately demonstrated in acceptance testing, and, since they are generally not time- or use-dependent, provide very little value to the customer in the postacceptance warranty period.

Next we will discuss the reliability, maintainability, supportability, and availability types of performance guarantees. These EPRs are desirable since they are time- and use-dependent, are measurable, and basically demonstrate the serviceability of the full-rate-production units in the field, over time. It is therefore the author's opinion that specification of EPRs and likewise the associated guarantees be limited to RAM-type parameters.

MTBF guarantee

Objective. The MTBF guarantee provides assurance that the required field MTBF level will be achieved. The MTBF guarantee provides a direct means for controlling the operational reliability of fielded systems.

Description. The contractor guarantees the field MTBF. Verification testing is conducted, and the results are compared with the guaranteed value. The contractor must develop and implement corrective action if the guaranteed MTBF is not achieved. Corrective action may also include provisions for consignment spares or downward price adjustment.

Two approaches have been used to determine guaranteed MTBF values (MTBFG): (1) the MTBFG value is specified in the request for proposal (RFP) or (2) contractors bid a MTBFG value. If contractors bid values, the RFP should at least specify a minimum value consistent with the system specification and development program. The bid value and the MTBFG price are potential source-selection factors.

An added consideration regarding MTBF guarantees is to allow for reliability growth over the warranty period. This can be accomplished by allowing a run-up or stabilization period where no MTBF values are measured. This period of time allows the contractor to identify and correct initial design, production, installation, and operation problems. A schedule pinpointing MTBF growth milestones can then be used to "grow" the MTBF to the desired value. From our RIW case study, the MTBF guarantee rider provided for three measurements over a 4-year period to grow to the desired level.

The contract must specify how MTBF is to be measured. If a current military data system is adequate to support such a measurement requirement, that data system may be used. In most cases, current data systems do not provide sufficient details or sufficient accuracy, and a controlled demonstration type of test is used. Creation of a unique data-collection process is time-consuming and costly and should be used only if the first two options are not feasible.

In the event a measured MTBF value fails to meet the guaranteed value, the contractor may be required to supply the following remedies

- Engineering analysis to determine the cause of nonconformance
- Corrective engineering design or production changes
- Modifications of systems, as required
- Pipeline consignment spares to support the logistics pipeline pending improvement in MTBF

With regard to the pipeline consignment spares remedy, typically a formula is used to determine the quantity of spares that reflects the shortfall in pipeline spares as a result of lower-than-expected MTBF. A maximum penalty is generally set to limit the contractor's liability. It is also possible to include an incentive if the MTBF exceeds the guaranteed value. The spares penalty is logical since the remedy, spares, directly helps the pipeline slowdown.

Applicability. MTBF must be the appropriate reliability parameter, and field measurement must be achievable to properly apply the MTBF guarantee. The MTBFG is best applied if the system is under contractor maintenance, that is, with an RIW to expedite the identification and remedy of problems. The system should be in production if the consignment spares remedy is invoked; otherwise this remedy is not practical.

Measurement. The measurement is specified in terms of measured relationship to the target or guaranteed MTBF. The following is the mathematical determination of the measured MTBF (MTBFM)

$$\text{MTBFM} = \frac{\text{total operating hours}}{\text{total relevant failures}} \tag{5.1}$$

It should be noted that if the warranted equipment's operating exposure is better represented by miles rather than time, the numerator becomes total operating miles and the measured parameter is mean miles between failures (MMBF). The numerator should include all operating hours, not just mission hours. Obviously there needs to be a mechanism for determining total hours, either elapsed-time indicators (ETIs) or by average mission length.

Logistics support-cost guarantee

Objective. The objective of the logistics support-cost (LSC) guarantee is to control and reduce LSC.

Description. The contractor bids a target LSC (TLSC) based on a model provided in the RFP. Key field parameters are measured, and the model is used to obtain the measured LSC (MLSC) by inserting the measured values of these key parameters in place of the target or guaranteed values in the contractor's proposal. The MLSC is compared with the TLSC. If the MLSC > TLSC, penalties in terms of free corrective actions or contract price adjustments are applied. If the MLSC < TLSC, an incentive such as an award fee may be applied.

The TLSC is usually defined through use of a model that combines acquisition costs, reliability, maintainability, and support factors. Cost elements included in a LSC guarantee are typically selected from the following cost categories:

- Hardware acquisition
- Initial spares
- Replenishment spares
- Organizational, intermediate, and depot maintenance
- Support equipment
- Support of support equipment
- Training
- Data
- Inventory management

Of these, the initial and replenishment spares and the maintenance categories are the most used in LSC guarantees.

The RFP should provide details on the model used to generate the TLSC. The algorithms should be clearly defined, and it should be personal computer (PC)-based to allow the contractor to conduct sensitivity analyses prior to setting the values for the guaranteed parameters to be measured in the field. The model should also include a set of standard factors such as organic maintenance labor rates and government shipping rates. The model needs to specify the number of operational systems and the length of the field use period. The guaranteed values must be contractor-controlled and are typically MTBF, MTTR, or hardware costs. Generally the contractor guarantees the TLSC and not the individually proposed parameter values unless explicit provisions are included for that purpose.

Several remedies are available with the LSC guarantee. One option is to use a contract price adjustment provision where the contract price is reduced by the amount proportional to the difference between the MLSC and TLSC, specifically, the support-cost overrun. Another option is to invoke a correction-of-defects clause in which the contractor must identify the causes of the overrun and then design and implement a fix. To provide positive incentives, there may be a provision that the contractor receives additional money if the MLSC is less than the TLSC. The calculation of the money due the contractor may be done by formula or through an award fee process.

Applicability. For proper application, an appropriate, well-defined model must exist. Also, the application of LSC guarantee may require a special test program to obtain valid measured values with which to compute MLSC. The LSC guarantee is used when the main focus for control is logistics support cost. The LSC guarantee has been used on such programs as the Air Force F-16 and the Navy F-18.

Measurement. The LSC measurement is based on operational evaluation testing focused on the use of the LSC model to determine compliance in terms of MLSC. Incentives or penalties are determined from the difference between MLSC and TLSC.

The following is a possible mathematical determination of LSC:

$$\text{LSC} = \text{CIS} + \text{CRS} + \text{CM} \tag{5.2}$$

where CIS = cost of initial spares
CRS = cost of replenishment spares
CM = cost of maintenance

CIS is a function of the equipment unit cost and MTBF, both contractor-controlled and measurable parameters. CRS is also a function of equipment unit cost and MTBF. CM is a function of equipment MTBF, and MTTR, where MTTR is contractor-controlled and measurable. Material cost is also measurable.

These three cost categories provide contractor-controlled and measurable parameters, which, as it turns out, are also typical support-cost drivers. Unless particular program requirements dictate otherwise, the author recommends that these three categories, at least, be included in the LSC model. It is recommended that the reader consult easily obtainable government LCC models such as the cost analysis strategy assessment (CASA) model to become familiar with the algorithms for these three and other applicable cost categories.

Availability guarantee

Objective. The objective of the availability guarantee is to ensure that required operational availability will be achieved.

Description. The availability guarantee focuses on measurable population characteristics. Typically, availability is specified as a threshold. Remedies include dollar penalties, no-cost units provided by the contractor, modification, redesign, or a combination of these in order to improve availability to the minimum specified level.

An availability guarantee is similar in concept to a MTBF guarantee in that it focuses on a measurable population characteristic rather than on individual system failures. In this case the characteristic is operational availability, or a measure of the system's readiness state.

Availability can be defined as the proportion of time a system is in a ready state (operational). In missile or ordinance applications, it could be percentage of units that are available for use, where the total number of systems is the denominator. There are several mathematical relationships for availability as follows:

$$A_I = \frac{\text{MTBF}}{\text{MTBF} + \text{MTTR}} \tag{5.3}$$

This form is called the *inherent availability.* It considers only the reliability and maintainability of the system in a perfect state, or independent of external influences such as maintenance and logistics delays. This relationship is useful in early design tradeoffs where the impacts of reliability and maintainability only are to be evaluated. Estimated MTTR is determined as

$$\text{MTTR} = \frac{\Sigma\lambda_i t_i}{\Sigma\lambda_i} \tag{5.4}$$

where λ_i is the failure rate of the ith unit and t_i is the estimated repair time for ith unit.

As it turns out, inherent availability is totally unrealistic when considering the availability of fielded units under a warranty. Availability must be a function of not only the capability to operate without failure, reliability, but also the capability to restore the system when it goes down. This restoration time, referred to as *mean downtime,* is a function of maintainability and logistics factors. What we really need is an operational availability relationship. We need to reflect the true downtime representing active maintenance problems and logistics delays. Equation (5.5) represents a relationship for operational availability as a function of MTBF and mean downtime (MDT).

$$A_o = \frac{\text{MTBF}}{\text{MTBF} + \text{MDT}} \tag{5.5}$$

It should be pointed out that MDT represents clock time and not worker-hours since we are representing the proportion of time that the system is operational. MDT is a function of administrative delays, logistics delays, and active repair time as follows:

$$\text{MDT} = \text{MADT} + \text{MLDT} + \text{MTTR} \tag{5.6}$$

where MADT is mean administrative delay time and MLDT represents mean logistics delay time.

The administrative delay time is the time from notification of a failure to the time the repair crew arrives at the failed equipment. In most cases we are concerned about the operational availability out at the use site, that is, the flightline. The impact of downtime is most critical at the use site since it hampers the accomplishment of the mission of that system. The logistics delay time is the time for securing the needed spare part(s) once the problem has been diagnosed. This time can be quite short or quite lengthy depending on whether a spare is available at the first-line repair facility. MLDT is then a function of spares availability and the spares turnaround time (TAT). Equation (5.7) shows the relationship for MLDT.

$$\text{MLDT} = \text{SAF} \times \text{TATSA} + \text{NSAP} \times \text{TATNSA} \tag{5.7}$$

where SAF = probability of having a spare available when needed
TATSA = turnaround time if spare is available
NSAP = probability of not having a spare available when needed (1 − SAF)
TATNSA = turnaround time if spare is not available

This relationship is used to estimate mean logistics delay time. In actuality, when the need arises, you either have the spare or you don't. The probabilities would be 1.0 or 0.0. This estimating tool for MLDT is useful when determining the likelihood of passing or not passing the availability requirement under the field warranty when the warranty is being quoted. From Eq. (5.7), if a large number of initial spares were procured, say, to provide a 95 percent assurance of having a spare when needed, for 95 percent of the failures, the TAT could be as short as an hour. If, however, a lower assurance factor were used for initial spares, then there would be a larger probability of not having a spare when needed and for a greater number of failures, the TAT could be as long as 24 h.

Some interesting tradeoffs arise relative to initial sparing. Laying in spares at a high level of assurance, as Eq. (5.7) shows, significantly reduces MLDT. A reduced MLDT reduces MDT [Eq. (5.6)], and reduced MDT raises A_o, [Eq. (5.5)]. High levels of A_o are attractive since that means that unit is available for use a higher percentage of the time, or a higher percentage of units are available for use. The other side of the tradeoff involves cost. It is expensive to lay in a large number of initial spares. Which is more important to that program, enhancing the A_o or spending less money on initial spares and using that money elsewhere in the program, say, for maintenance training? This can be a tough call and requires a thorough look.

To illustrate the impact of sparing on MLDT, let's assume that for the item under warranty a 95 percent spares assurance factor is thought to be needed, and if a spare is available, the average TAT is 1.0 h. If the spare is not available, the average TAT is 24.0 h. MLDT is calculated as

$$\begin{aligned}\text{MLDT} &= 0.95(1.0) + 0.05(24.0)\\ &= 0.95 + 1.2\\ &= 2.15\text{ h}\end{aligned}$$

MLDT is the weighted average delay time based on whether a spare is available when needed. For our problem, the weighted average logistics delay time is 2.15 h. This is the number to be used in A_o estimates to evaluate the risk in passing the availability requirement. As a point of comparison, if the spares availability factor drops to .90, then MLDT is calculated to be 3.30 h. A 5 percent drop in SAF increases MLDT by 53.5 percent. These types of evaluations are essential to the thoroughness of risk evaluations for availability validation, and, in a larger sense, the total warranty quote.

Referring back to Eq.(5.5), MTBF may not be the most realistic measure of average time between downing events. MTBF, as its name communicates, is the average time between failures or legitimate removals. But what of other operations that could put the system down such as checks and adjustments? These, too, must be comprehended in the evaluation of operational availability. A more realistic reliability measure could be mean time between maintenance (MTBM). MTBM captures all the activities that put the system down, including scheduled maintenance. Obviously, MTBM < MTBF and makes MDT in Eq. (5.5) a more influential factor in the calculation of A_o. Considering MTBM as the reliability measure changes Eq. (5.5) to

$$A_o = \frac{\text{MTBM}}{\text{MTBM} + \text{MDT}} \tag{5.8}$$

Equation (5.8) provides a more thorough measure of A_o for warranty quoting purposes. Relative to the A_o measurement, if the customer is going to use a calculated value of A_o from the demonstrated reliability, maintainability, and logistics factors, the contractor's MTBF (MTBM) projections must be based on field conditions. To do this, the predicted, or inherent reliability must be factored to account for improper handling in the field and induced problems. For example, the contractor may want to divide the predicted reliability by a number from 2 to 5 to arrive at the true periodicity of downing events in the field. The 2-to-5 range has been demonstrated on systems in the field. Whether using MTBM or field MTBF, the contractor's concern should be how often the equipment will be down and what impact that has on the demonstrated A_o.

It is interesting to note that some commercial companies whose product's availability is critical to safety or medical operations, guarantee the total downtime rather than A_o. For instance, they may guarantee that their system will be down only 6 h per month, including scheduled maintenance. They report that total downtime per unit time is much more understandable and pragmatic to the customer than operational availability. The bottom line would be to know your customer and deal with the availability guarantee in the terms your customer understands, and gear your analyses thusly.

In practice, an availability guarantee is implemented in a manner similar to an MTBF guarantee. Availability values are specified in the contract. Periodic measurements of fielded systems are made to obtain operational availability statistics. If the measured operational availability is less than the contractually guaranteed value, warranty remedies are invoked.

It is necessary for the customer to recognize that the downtime component of availability may involve elements that are not under contractor control, such as logistics delay times waiting for spares or tools. The guaranteed value and corresponding measurement procedure should not penalize contractors for negative factors over which they have no control and therefore are not liable for. Perhaps the customer could provide "standard" values for the factors not under contractor control which could be combined with MTBF (MTBM)- and MTTR-controllable factors to use for the contractor's quote. The measurement would be of the contractor-controlled values, and A_o calculated using those values to determine compliance. A true A_o would not be calculated, but contractor compliance could be determined.

Applicability. An availability guarantee is most applicable for systems that are dormant for an extended period of time such as missile systems or systems that must operate continuously such as airport surveillance radars. Despite the extended periods of dormancy in storage, missiles typically have a high operational availability requirement. Because of the safety aspect, airport surveillance radars also are required to demonstrate very high levels of operational availability, sometimes as high as four 9s, (i.e., 99.99 percent) or even higher in some instances.

Measurement. For dormant systems, data from periodic checkouts, test launches, built-in test equipment (BITE) checks, and other special checks can be used to determine the percentage of good units, based on a sample of units from storage. For continuously operating systems, the ratio of uptime to total time may be measured, or individual measurements of MTBF (MTBM) and MTTR could be combined to provide the needed availability statistics.

Advantages and Disadvantages

Looking at performance guarantees as a whole, the major advantage is the assurance that the achievement of required levels relative to the essential or key performance parameters, as well as the warranted item's repair, will be the contractor's responsibility. The achievement of these performance levels is critical since it gives the customer assurance that the mission of the warranted item can be accomplished with full-rate production fielded systems and all the misuse and wear that go with it. With only repair and or redesign coverage, this detailed assurance would not be established.

The major disadvantage would be the added administrative complexity to the warranty program. The up-front warranty requirements would be more complex and time-consuming to prepare, the negotiations would be lengthier, and, most importantly, the implementation of the warranty would be considerably more involved than a repair warranty. The data collection and analysis related to performance guarantees can be quite costly and involved, and may even require the development of a new tailored data system. Use of an existing data system can involve hours of laborious interpretation for both customer and contractor. Finally, the documentation and tracking of the results to determine whether penalties or incentives are invoked and the necessary follow-up to ensure that required actions are taken can also add to the administrative complexity.

Criteria for Use

For weapon systems that meet the criteria of Title 10, Sec. 2403, the statute requires performance guarantees to satisfy the essential performance coverage under the law. For commercial suppliers, the decision to apply performance guarantees is much tougher since they are offered voluntarily. They have the option to offer no performance guarantees or to offer certain selected ones to be used as marketing factors to capture new or retain current customers based on the competition.

Whether performance guarantees are required or optional, it is incumbent on the government customer or commercial supplier to carefully evaluate the advantages and disadvantages of performance guarantees and choose the ones that provide the best overall value to the customer. Value is measured in terms of what added quality assurance the customer gets for the added cost.

Examples

Following are examples of actual performance guarantee requirements experienced by the author:

- LSC guarantee
  - LSC covered initial spares and maintenance cost categories

  - Target LSC (TLSC) in contractor's proposal
  - Measured LSC (MLSC) from the field-test program
  - TLSC compared with MLSC
  - MLSC ≤ TLSC—success
  - MLSC > TLSC by >25 percent—seller does free corrective action
  - No incentives
- MTBF guarantee
  - Guaranteed MTBF (MTBFG) compared to demonstrated MTBF (MTBFD)—test program
  - MTBFD < 85 percent of MTBFG—seller provides free additional spares
  - No incentives
- Turnaround time (TAT) guarantee
  - Measured and required TATs compared—contractor data
  - TATM > TATR—seller provides free spares
  - No incentives
- Built-in test (BIT) testability guarantee
  - Requirements

    Contract items shall achieve false-failure indication (FFI) of < 2 percent

    Contract items shall achieve fault detection (FD) of ≥ 95 percent
  - Penalties if requirements not met in controlled test

    Perform engineering analysis to determine cause of nonconformance at no additional cost to government

    Prepare no-cost ECPs based on engineering analysis

    Implement no-cost ECPs

    Furnish additional pipeline spares per established algorithm at no additional cost to government
  - No incentives
- System mission readiness (SMR) guarantee
  - Missile system
  - Requirement

    SMRR ≥ 90 percent

    Seller penalized downtime hours for failure to restore system in specified time

    Downtime hours degrade SMRM
  - Penalty—SMRM < 90 percent—seller pays $1000 per percentage point below 90 percent per system per month
  - No floor on SMRM to limit contractor's liability
  - No incentives
- Shop visit rate (SVR) guarantee
  - Aircraft engine
  - If actual SVR exceeds the contract limit, contractor pays the government a penalty based on amount above the limit

- If actual SVR is lower than contract limit, the government pays the contractor an incentive based on amount below the limit
- Deadband zone exists between the upper and lower SVR limits

In this example, the SVR requirement value was not documented in the warranty requirement.

To illustrate how the performance guarantees are integrated into the overall warranty requirement, the following is a warranty statement prepared by a contractor in response to the government's request for the contractor's "best" warranty. To retain anonymity, the contractor's name, the customer's name, and the equipment identification are omitted, hence the (XXXXX).

1.0 PRODUCTION WARRANTY

1.1 *General*

(XXXXX) is firmly committed to customer satisfaction and the implicit considerations of performance and quality. We understand the warranty legislation and intend to offer a warranty that adheres to the law and provides (XXXXX) with a flexible, cost-effective program within the boundaries of the maintenance/support plan.

1.2 *Commitment*

1.2.1 *Conditions*

(XXXXX) warrants that the (XXXXX) furnished under this contract:

i) at the time of acceptance, shall conform to the design and manufacturing requirements in the contract,

ii) at the time of acceptance, shall be free from all defects in material and workmanship, and,

iii) shall conform to the following Essential Performance Requirements (EPR):

- Mean Time Between Failures (MTBF) = TBD hours,
- Contractor Depot Mean Turnaround Time (CDMTAT) = TBD days, maximum = TBD days.

1.2.2 *Duration*

i) 24 months or 432 A/C flying hours, whichever occurs first, after (XXXXX) acceptance (DD-250) for each (XXXXX) initially delivered for repair/replacement of all applicable failures,

ii) for 36 months after acceptance of the first initially delivered (XXXXX) for correction of defects,

iii) for 36 months after acceptance of the first initially delivered (XXXXX) for MTBF and CDMTAT penalties and,

iv) corrected items assume remaining warranty of initially delivered item.

1.2.3 *Remedies*

In the event of a (XXXXX) verified relevant breach of the warranty in paragraphs 1.2.1 and 1.2.2 above, (XXXXX) direction, (XXXXX) will take the appropriate actions to correct the breach at no increase in cost to (XXXXX), including:

i) within the average of TBD days from the date of receipt of failed supplies at (XXXXX) plant, (XXXXX) shall complete all appropriate corrective actions, including repair or replacement, and tender the corrected supplies for redelivery to (XXXXX). A failure is any malfunction of an (XXXXX) which renders it unserviceable.

ii) within 60 days of written notification from (XXXXX) of defective supplies, (XXXXX) shall submit to the contracting officer a written plan with recommended actions and associated schedule to remedy the defect. The plan will include analysis of defect, proposed corrective action, and a redesign/retrofit strategy, if directed by (XXXXX). A defect is defined as a breach of the MTBF requirement.

iii) within 45 days of submittal of the defect correction plan, (XXXXX) will approve/disapprove the plan; no response from within 45 days will be interpreted by (XXXXX) to constitute approval.

iv) provide additional depot spares if the combined (XXXXX) CDMTAT and MTBF performance during a measurement period results in the need for additional depot spares assets over that which currently exists in the inventory for that measurement period. The details of the CDMTAT and MTBF measurements and penalty spares calculations are discussed in 1.3, Warranty Administration.

v) prepare and furnish new/revised data and reports related to corrective actions per the DD 1423s.

1.2.4 *Transportation*

i) (XXXXX) bears the cost of transportation of failed supplies to (XXXXX) plant.

ii) (XXXXX) bears the cost of transportation of failed supplies back to (XXXXX).

1.2.5 *Exclusions*

Failures and defects resulting from the following conditions are considered nonrelevant and are not covered under this warranty:

i) improper handling by (XXXXX) or (XXXXX)

ii) (XXXXX) or (XXXXX) maintenance not in accordance with approved TO's,

iii) combat damage, and

iv) catastrophic events—fire, flood, crash, and other uncontrollable events.

1.2.6 *Markings*

i) for each warranted depot repairable item, (XXXXX) shall provide complete, accurate, and legible warranty information as part of the acceptance of each initially delivered item.

ii) all depot repairable warranted items under this contract shall be identified by (XXXXX) by marking each item per MIL-STD-130.

iii) all warranty markings shall be indelible, legible, and include the following information:

- "WARRANTED ITEM" in large bold letters
- NSN, part no., or serial number/item identifier,
- contract no.,

- warranty expiration data, and
- "APPROVED ORGANIC REPAIR WILL NOT VOID THE WARRANTY"

1.3 *Warranty Administration*

(XXXXX) will provide flying hour data for each warranted, (XXXXX) and (XXXXX) will provide monthly depot turnaround time data. Following are the details relative to warranty collection and measurement for the essential performance requirements:

i) monthly failure and flying hour data shall be collected and accumulated on all (XXXXX) beginning with operation of first (XXXXX), and total depot turnaround days shall be collected monthly for depot repairable items, beginning with operation of the first (XXXXX).

ii) MTBF shall be calculated as total reported (XXXXX) flying hours × 3.2 divided by total number of verified relevant failures. Measurement of MTBF shall begin after total equipment flying hours have accumulated equal to twice the required MTBF hours. The MTBF achieved at the first measurement point shall be 50% of that guaranteed. Calculation and documentation of MTBF shall continue at six month intervals for purposes of tracking and voluntary corrective action. The second measurement point shall be when total equipment flying hours have accumulated equal to eight times the required MTBF hours. The MTBF achieved at the second measurement point, and beyond, shall be 100% of that guaranteed. Measurement continues at six-month intervals until 36 months after operation of the first (XXXXX).

iii) average depot turnaround time for repair (CDMTAT) shall be measured coincident with the second MTBF measurement using the accumulated data from the operation of the first (XXXXX), and every six months thereafter using only that six months worth of data, until 36 months after operation of the first (XXXXX) as:

$$\text{CDMTAT} = \frac{\text{TAT days at depot for relevant failures for period}}{\text{No. of relevant failures for period}}$$

iv) penalty spares, as appropriate, shall be calculated coincident with MTBF and CDMTAT measurements using the calculated data in ii) and iii) above as:

- If $\text{CDMTAT}_M \le \text{CDMTAT}_R$ and $\text{MTBF}_M \ge \text{MTBF}_R$—no penalty spares
- If $\text{CDMTAT}_M > \text{CDMTAT}_R$ or $\text{MTBF}_M < \text{MTBF}_R$—possible penalty spares as follows:

  For each depot repairable item:
  Qty. of Spares = Expected μ + Assurance $(Z\sqrt{\mu})$

$$\text{IDS} = \frac{\text{Tot. negotiated operating hrs. per day}}{\text{MTBF}_R} \times \text{CDMTAT}_R + Z\sqrt{\mu}$$

$$\text{MDS} = \frac{\text{Tot. negotiated operating hrs. per day}}{\text{MTBF}_M} \times \text{CDMTAT}_M + Z\sqrt{\mu}$$

Where:
$MTBF_R$ and $CDMTAT_R$ are required values, and
$MTBF_M$ and $CDMTAT_M$ are measured values.
Z – standard normal variate for 90% level of spares assurance, and, IDS and MDS are initial depot spares and measured depot spares quantities, respectively.

If MDS ≤ MAX (IDS, PREV. PERIODS' MDS)—no penalty spares
If MDS > MAX (IDS, PREV. PERIODS' MDS)—number of penalty spares (NPS) as follows:
NPS = MDS – MAX (IDS, PREV. PERIODS' MDS)—for each applicable depot repairable item.
Initial depot spares will be included in (XXXXX) quote to (XXXXX) based on a 90 percent assurance factor and meeting MTBF and CDMTAT requirements.

1.4 *Warranty Review Board (WRB)*

(XXXXX) will host the WRB meetings, the first meeting being held coincident with the first MTBF and CDMTAT measurement, and every six months thereafter. (XXXXX) will chair the WRB. (XXXXX) will assist in the judgment of the relevancy of (XXXXX) flying hour and failure data since the previous meeting, and the calculations of (XXXXX). MTBF and CDMTAT, and penalty spares, if any, for the previous period, per 1.3 above. This will be done for the purposes of identifying any warranty breaches leading to redesign/retrofit and/or spares penalty remedies, as appropriate.

1.5 *Adjustments*

i) (XXXXX) will be entitled to an adjustment equal to the negotiated cost of repair/replacement for any corrected failures at depot later found to be nonrelevant under the warranty.
ii) (XXXXX) shall be entitled to an adjustment equal to the negotiated cost of repair/replacement of the depot repairable item if they choose not to have the failure repaired/replaced.
iii) (XXXXX) or (XXXXX) are entitled, respectively, to an adjustment equal to the number of failures incurred by more operating hours than that negotiated times an average cost of repair, where the required MTBF is used, or equal to the number of failures avoided by less operating hours than that negotiated times an average cost of repair, where the required MTBF is used. These judgments and calculations will be performed at the WRB meetings.

With respect to this example, note in Sec. 1.2.1(iii), that the essential performance parameters are defined to be MTBF and CDMTAT. The MTBF requirement is for the total warranted system. The to-be-determined (TBD) notation was used in place of the actual values, again, for anonymity purposes. CDMTAT is a dock-to-dock requirement and contains no time intervals out of the contractor's control.

In Sec. 1.2.2(iii) the duration of the EPR coverage is given to be 36 months after government acceptance of the first initially delivered item.

Sec. 1.2.3(iv) specifies the remedy for nonconformance to the combined CDMTAT and MTBF performance guarantees during a measurement period. The penalty is providing free spares to the government if the spares requirement for that period exceeded what was currently existing in the spares inventory. Since both MTBF and CDMTAT affect the availability of spares assets when not met, the decision was made to use their combined effect, rather than one or both of them being noncompliant in any measurement period. This approach represents less risk to the contractor.

Sec. 1.3(ii) pinpoints when and how the MTBF is to be determined. Notice an MTBF growth program is proposed, allowing the contractor time to remove inherent problems before 100 percent of the MTBF requirement is imposed; also allowing a 2 × MTBF-h run-up period before MTBF is measured. The measurement of MTBF is, as described in this chapter's section on MTBF guarantees, the quotient of operating hours divided by the total number of verified relevant failures.

The specifics of the CDMTAT performance guarantee are described in Sec. 1.3(iii). The CDMTAT measurement statistic is defined to be the quotient of total TAT days at depot for relevant failures for a measurement period divided by the number of relevant failures in that period. This data is kept by the contractor, available for government scrutiny. Note that this is an average TAT and not a requirement on each and every item that comes back to depot for repair. Here again, this approach involves less risk to the contractor. The measurement intervals are coincidental with that of MTBF, beginning with the second measurement.

Sec.1.3(iv) provides the criteria for penalty spares, as applicable. The equations are clearly documented. First, the criterion for no penalty spares is shown, that is, both measured MTBF and CDMTAT meet or beat the requirement. If either MTBF or CDMTAT do not meet the requirements, penalty spares may be required. A general spares algorithm is used to calculate initial depot spares and measured depot spares using the guaranteed or required values and the measured values of MTBF and CDMTAT, respectively. If the measured value of depot spares for that period is less than or equal to the higher of the initial or previous period's spares levels for that particular depot repairable item, then no penalty spares are required. If the measured value of depot spares for that period is greater than the higher of the initial or previous period's spares levels for that particular depot repairable item, then the number of penalty spares is computed as the difference between the two values. A 90 percent spares assurance factor is used in the initial and measured spares calculations.

Finally, in Sec. 1.4, the Warranty Review Board's responsibilities are outlined. Both contractor and customer will be involved in judging the relevancy of flying-hour and failure data since the last meeting and in the calculation of MTBF and CDMTAT and penalty spares, if any, for the previous measurement period. The board's meetings will be held coincident with the first MTBF and CDMTAT measurement, and every 6 months thereafter.

Referring back to the "Definition" section of this chapter, we have verified that, in the previous example, the essential performance parameters were clearly specified as to

- Which ones are measured
- How are they measured
- When are they measured
- Criteria for compliance or noncompliance
- Method of determining the penalties for noncompliance

Additionally, the EPRs chosen were either reliability- or supportability-type parameters, giving continuous indication of the serviceability of full-rate production units in the field over an extended period of time.

It is possible to write warranty requirements that are clear, concise, and thorough with respect to EPRs and their validation via performance guarantees. While the previous example may have seemed to contain excessive detail relative to the performance guarantees, this level of detail is essential to the realistic quoting and implementation of cost-effective warranties for both sides.

We, by no means, have covered all possible performance guarantees in our examples. There are, no doubt, numerous commercial type performance guarantees such as supplier response time guarantees on such things as home security systems which we have not addressed.

One final word on performance guarantees, lest we minimize their importance. A supplier of commercial systems was reported to have flown a spare part to a location outside the United States to avoid incurring a penalty on an operational availability performance guarantee. In my judgment, the penalties must have been extremely severe to necessitate the use of a company jet to fly a spare part to that location to render the system operational. Enough said.

Chapter 6

Warranty Tradeoffs

General Considerations

In this chapter we will discuss those factors that the seller should consider before formulating a warranty statement. The discussion is applicable to all business sectors. Because warranty requirements from the customer rarely exist in the consumer and commercial sectors and tradeoffs are crucial to establishing the warranty policy, we will discuss warranty tradeoffs in the consumer and commercial contexts. These analyses are performed from business and statistical viewpoints and provide the rationale to support the items which should be stated in the written warranty.

Warranty Analysis Considerations

Before offering a warranty, the seller must understand:

- Performance and serviceability aspects of the product
- Customer marketplace
- Legal issues

The remaining discussion of warranty analyses will center on the performance and serviceability aspects of the product. It is assumed that extensive market research is done up front to understand the user's needs and operating environment. It is further assumed that appropriate legal counsel is consulted to make judgments and recommendations on the legal issues.

We perform a warranty analysis because we need to know how much the warranty service will cost. More specifically, we would like to know the following:

- The expected number of free replacements
- The expected cost to the buyer
- The expected profit to the manufacturer

The key warranty analysis steps are as follows:

- Estimate the life distribution of the product
- Understand the various types of warranty policies
- Develop models that permit the required tradeoff analyses

An important footnote to these three key analysis steps is to remember that what is not stated in a warranty may be more critical than what is stated.

Figure 6.1 shows the major disciplines involved in a successful warranty analysis.

Note that Fig. 6.1 is actually an equation. All the disciplines shown must be present before an informal warranty policy can be established. Market research begins as the product concept is being formulated. Initial reliability analyses can begin with data from similar equipment. Models are refined as data are acquired for the product being manufactured.

Warranty analyses require the following:

- Performance data from similar earlier equipment and from the product being manufactured
- Extensive in-house testing on critical subsystems, engineering models, and early production models

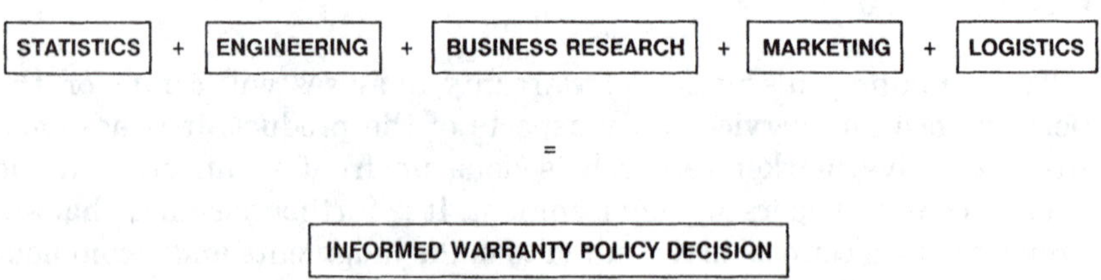

Figure 6.1 Major disciplines in a commercial warranty analysis.

- External trade testing at customer sites for reliability analysis and immediate feedback of early-life problems into the corrective action system
- Extensive market research to understand the user's operating scenario
- Benchmarking with competitors' products
- Tradeoffs among the different service strategies relative to site location, staffing, and parts inventories

After performing a warranty analysis, the manufacturer can do the following:

- Quantify the service requirements of the product over time
- Understand the service strategy necessary to provide warranty support expected by the buyers
- Understand the warranty strategies of competitors

With this knowledge, the associated tradeoffs, and the help of legal counsel, a formal warranty statement can be drafted.

As a caution, all data are subject to variability. Results of analyses should have statements of the associated uncertainties. Also, more than one scenario should be studied, and sensitivities to the many assumptions should be investigated.

Tradeoff Issues

Warranty tradeoff analyses help the seller formulate a warranty strategy based on facts. Several factors must be studied and decisions made regarding their relative importance in an optimal warranty strategy. The factors that can be traded off by analysis are as follows:

- Length of warranty
- Type of warranty policy
- Breadth of warranty coverage
- Type of remedy
- Service strategy
- Administrative costs
- Competitors' warranties

These factors will be discussed as follows:

Length of Warranty

The length of the warranty should reflect the performance of the product. It is not wise to set the warranty duration at a usage value associated with a high failure percentage of the product since too many claims would be possible. The life of the product needs to be understood in a statistical sense. Data can be gathered from a variety of sources such as subsystem and system testing, external trade trials, and historical databases of similar equipment. Appropriate meters need to be incorporated into the equipment design to record the life of the product. It is also important to obtain failure causes for each event.

The product being warranted will either be repairable or nonrepairable. The appropriate analysis for each category will be different.

Nonrepairable items. A *nonrepairable item* may be the total product or a separately warranted module. When analyzing data from a nonrepairable item, it should be verified that the data are from a homogeneous population, and that there have been no design or material changes which would cause one item to last longer than another.

When studying a nonrepairable module in a repairable system, one should verify that the arrival pattern of the failures of these modules in the system does not exhibit either a trend toward decreasing times between failures or increasing times between failures. A random arrival is likely to occur when a failed module is replaced with an item of the quality of the original one. If this is the case, then it would be reasonable to assume that the module was exhibiting a renewal process in which the times between successive module replacements were independent and identically distributed with an unknown distribution.

The underlying life distribution for the product can be estimated using reliability models based on the Weibull or lognormal distribution. Goodness-of-fit tests should be applied to verify model validity. Confidence intervals should be stated for the parameter estimates. The expected number of renewals in a given time period can then be calculated assuming an ordinary renewal process starting at time = 0.

Repairable items. The methodology for analyzing repairable systems is more complex than for nonrepairable systems. Three analysis paths are described.

A trend test should be applied to the failure arrival data for each studied product in the sample. If the arrival pattern does not exhibit a trend toward decreasing times between failures or increasing times between failures, then it can be assumed that the equipment as a whole is exhibiting a renewal process. The entire product can be studied as if it were nonrepairable using the methodology described above.

If there is a trend, then the entire system can be studied using graphical repairable system methodology. Two such methods are the rate of occurrence of failures (ROCOF) and the mean cumulative number of failures (MCNF). The natural estimate of the ROCOF for the sample may be found by assuming that the underlying ROCOF for each equipment in the sample is similar and can be pooled. A graph of ROCOF versus time will provide information about the improvement or deterioration of the system. The nonparametric graphical estimate of MCNF or mean cumulative cost can be used to evaluate whether a cost or failure frequency increases with system age. For each method, confidence intervals on the estimates can be obtained using a bootstrap method.

Finally, if there is a trend and it is desired to do a detailed system analysis, data need to be obtained for all major subsystems and components. Data on these individual modules can be fitted using appropriate reliability models. Then the overall system can be modeled using a series-parallel (redundancy) combination of the modules. To be thorough, the model should take into account such factors as the uncertainty in the distribution parameters' estimates, the concept of incomplete or new-better-than-used repair, and the aging of parts in the system. This is best handled in a simulation model.

It should also be noted that the initial installs of a new product have the potential for more problems than later installs. Figure 6.2 shows the analysis process related to the length of the warranty as described above.

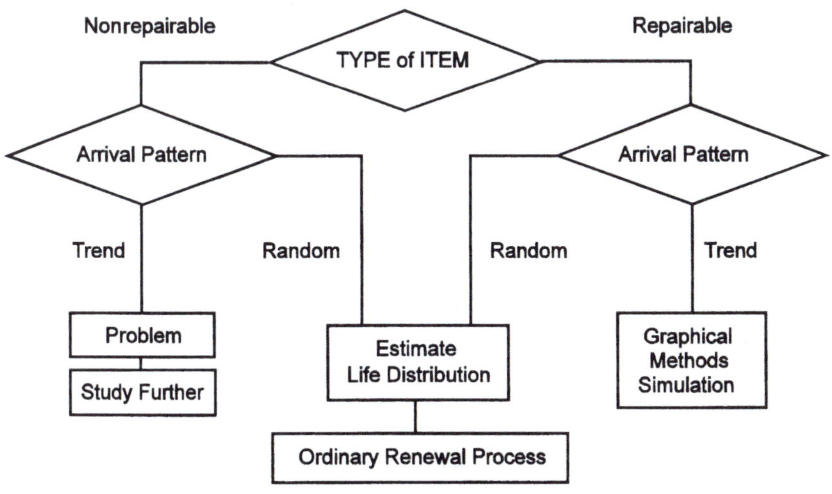

Figure 6.2 Commercial warranty analysis process.

Types of warranty policies (nonlegal sense)

Knowledge of the product life distribution leads to understanding of the expected number of failures and/or remedies during a stated time period. A study of the types of warranty policies and associated cost models should lead the seller toward a warranty strategy which moderates incurred expenses. There are several types of warranty policies. Four commonly used policies will be discussed here.

Free replacement warranty policy. Under a free repair or replacement policy, the seller pays the entire cost of the remedy if the product fails before the end of the warranted period. Thus, for long coverage times the warranty costs can be very large, and the number of replacement purchases over the product life cycle will be reduced, which in turn reduces the total profits.

In an ordinary free replacement policy, the remedied item has a warranty equal to the remaining length of the original warranty. Such a warranty assures that the buyer will receive as many free repairs or replacements as needed during the length of the original warranty. This type of policy is commonly used for consumer durables such as automobiles.

With an unlimited free replacement policy, the replacement item carries a warranty identical to the warranty on the original purchase. Thus the warranty is renewed with each replacement. This policy is often used for small electronic appliances which suffer from infant mortality and is usually limited to very short time periods. The required bookkeeping is less than that for the ordinary free replacement warranty policy. The free replacement policy favors the buyer at the expense of the seller.

Pro rata warranty policy. Under a pro rata warranty, if the product fails before the end of the warranty period, it is replaced at a cost which depends on the age or wear of the item at the time of failure. Typically, a discount proportional to the remaining length of the warranty is given on the purchase price of the replacement item. The replacement item usually carries a warranty with terms identical to those on the original product.

The pro rata warranty policy is most often used for items which wear out and must be replaced at failure, rather than items which may be repaired. Typical examples are vehicle tires and batteries. A pro rata policy is more appealing to the seller but unattractive to the buyer since the buyer may have to purchase a new item, at a cost, should the earlier item have a very short useful life. Thus, the pro rata structure favors the seller at the expense of the buyer.

Combination policy. A combination policy contains both free and pro rata periods. This policy has an initial free replacement period followed by a pro rata period during which the cost of the replacement item is calculated on a sliding scale. The relative lengths of the free and pro rata periods for the same overall warranty period can be studied on a cost basis.

In the combination policy, there is a consideration of whether the warranty is renewed only after each purchase or after each failure. Formulas for both buyer's costs and manufacturer's profits have been derived under the assumption of a renewal process using the results of a renewal reward process for the long-run average costs.

A combination policy has a promotional appeal to attract buyers and at the same time keeps the warranty costs for the seller within a reasonable amount.

Fleet warranty. A fleet warranty covers a population of items and therefore may be appropriate when a large number of identical items is being sold to a common buyer with the understanding that replacement parts will be provided by the manufacturer. In a fleet warranty, the manufacturer guarantees that the mean life of a population of items will meet or exceed some negotiated mean. If the mean fleet life meets or exceeds the warranted value, then no compensation is given by the manufacturer, even if individual items have very short lifetimes. If the mean life is less than the guaranteed value, compensation is given according to how much the specified mean life exceeds the observed mean.

There has been little published on fleet warranties. Recent work has focused on nonrepairable items within more complex systems. Critical to the fleet warranty are the estimation method for the mean, the length of the time window, and the starting point for the window (ordinary versus equilibrium renewal process).

Breadth of warranty coverage

The manufacturer usually assumes responsibility for defects in materials and workmanship. However, the entire product need not be covered. Warranties seldom include consumables such as filters and software diskettes, peripherals and interfaces not purchased from the manufacturer. Repair, maintenance, alteration, and/or modification of the product by other than manufacturer-authorized personnel are often reasons for voiding a warranty.

Type of remedy

Failed items are typically repaired or replaced. The repair option is based on such factors as item complexity, original cost, and ease and efficiency of repair. Circuit boards are often repaired and placed back into spare-part inventory. Modules may be fabricated from rebuilt parts. The buyer must be notified in the warranty if used parts might be employed in a repair. Environmental concerns are promoting material recycling if repair is not feasible. Under a full consumer warranty, the buyer is entitled to a refund, less depreciation, if the defect cannot be remedied within a reasonable number of attempts.

Service strategy

The manufacturer needs to understand the buyer's operating environment when setting a service strategy. Some questions which should be answered are

- Is the product's function critical to the buyer's business?
- Will the product be used 24 h per day, 7 days a week, or from 8 to 5 Monday through Friday, or on a casual basis?
- Is the buyer willing to do minor maintenance?
- Is the product designed so that it is easily serviced by a person with minimal training?
- If the equipment is portable, is the buyer willing to send it back to the manufacturer or take it to a service center? How far are they willing to drive?

There are many possible service strategies, and each carries with it many logistics issues. Common scenarios are

- *Mail-in return.* Responsibility for the cost of packaging and shipping, TAT on the repair.
- *Walk-in service center.* Staffed by manufacturer-seller or authorized personnel, hours of service.
- *On-site service.* Provided by manufacturer-seller or authorized personnel, hours of operations, fees for off-hours service, field replaceable modules.
- *Telephone assistance.* Provided by manufacturer-seller or authorized personnel, level of training required, hours of service.

Telephone assistance is often a first step in screening the need for a site visit to remedy a problem. The use of remote diagnostics for complex equipment has reduced the need for on-site assistance and has shortened the call duration for those events which do require a site visit.

Administrative costs

All warranty plans have administrative costs. Records must be kept indicating the start of each item's warranty. For consumer products, the buyer is often responsible for keeping the sales slip. Databases detailing the defect type and location, the defect remedy, and the amount of usage at the time of the problem, must be built, maintained, and analyzed.

Competitors' warranties

A product's warranty must be competitive with those stated for similar products. If the optimal warranty policy for your product is broader than what is being offered by the competition, then your warranty could become a selling point. If this is not the case, then you must reanalyze your product using the warranty scenario set forth in the competitors' strategy. The cost implications should be studied, and areas for design improvement identified.

Tradeoff Process

Figure 6.3 shows the overall tradeoff analysis process.

Basic to any tradeoff analysis is the definition of the purpose of the analysis. For example, we may wish to know which is the best warranty duration for a product, given that the other tradeoff factors have been evaluated and the optimum levels set. The candidates are then identified. All identified candidates must be viable; that is to say, they must make sense in the context of competition and must be implementable. If the competition is offering a 12-month warranty, then a 36-month or a 6-month warranty does not make good business sense in the context of competition even though either is implementable. In our example at the end of the chapter, the two candidate durations might be 12 and 18 months.

What is the basis for comparison or the figure of merit to be used to rank the alternatives? Is it cost, reliability, or availability? In our example, the figure of merit could be the ability to compete while holding the profit line. This is an extremely important step in the tradeoff process. Many well-intentioned engineers have spent much time developing and performing a tradeoff analysis only to learn at the end that

Figure 6.3 Warranty tradeoff process.

their chosen figure of merit did not adequately discriminate between or among the candidates; therefore, a meaningful ranking of the candidates was not achievable.

The model development should be geared to the availability of input data to support it. It is of little value to have a very detailed set of algorithms in the model and having insufficient data definition at that point in time to support it. The complexity of the mathematical-statistical model should be directly proportional to the level of definition of the input data. Since we are interested in comparing the figures of merit of the candidates to yield a ranking, it is necessary to consider elements in the model only where differences are expected to exist between or among alternatives. It is essential to develop these models to run on a PC using Lotus or other spreadsheet software, especially if simulation is the method of analysis.

The next step is singularly the most vital in the entire process. Since most models are evaluators, not optimizers, the step of input data estimation is crucial to the evaluation of the candidates. If the input data does not reflect the differences among the alternatives, then the model outputs cannot reflect the differences. To secure reasonable estimates for input data, a team of knowledgeable people representing the necessary disciplines is required. For example, the reliability engineer on the team must be knowledgeable not only in the reliability discipline but also relative to the product itself. These criteria apply to all members of the team.

One of the team members' responsibilities is to document the rationale for their estimates. It is important to know whence the estimate came, especially when the ranking is quite close and a further examination of the input data is required and possible reruns made. The care with which the input data is estimated can make or break the validity of the analysis. A more detailed discussion of the warranty team relative to structure and responsibilities is found in Chap. 11.

Having developed the model and secured the input data, we can now perform the baseline analysis. The importance of the baseline analysis cannot be overemphasized. An initial stake in the ground must be established. This stake is the baseline analysis. The baseline analysis provides a frame of reference from which to establish the sensitivities of the driving input parameters. The baseline analysis represents the most likely results based on the most reasonable inputs from the team. The initial baseline remains the baseline until the design and/or support considerations of the product change through tradeoff analyses and a new baseline is established.

With the results of the baseline analysis in hand, the ranking of candidates can be achieved. The figure of merit established in our third step should clearly indicate whether high is good or low is good, and the candidates can be ranked accordingly. In interpreting the results, care should be taken to determine the percentage difference(s) between or among the figures of merit for the candidates. This will be especially crucial in the final decision step.

With the results of the sensitivity analysis performed in conjunction with the baseline analysis, we can perform sensitivity analysis with regard to the tradeoff analysis results—the initial ranking. By varying the driver parameters from the baseline analysis plus and minus—say, 25 percent—and rerunning the tradeoff analysis, we can see the effect on the ranking. This can be accomplished as a univariate, or multivariate, operation, the ultimate being a simulation approach. What we have done is identify the thresholds on the input data drivers where the candidate rankings change. Because of the estimative nature of the total process, sensitivity analysis is essential to help us determine under which conditions candidate A is better, under which conditions candidate B is better, and so on. Rarely does one candidate always prevail under all values of the input data drivers.

Finally, the final step—the decision process. To select the best candidate, we must study the results of the tradeoff sensitivities that led to the final ranking. If possible, a plot of the relative ranking as a function of the driver parameter values should be made. In this way a judgment can be made as to what region of the plot is most likely to occur with respect to the input driver value. To illustrate, Fig. 6.4 shows the re-

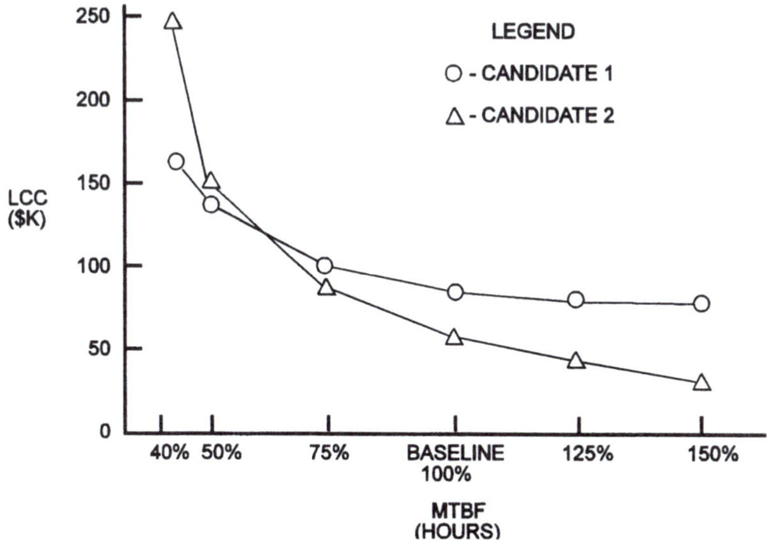

Figure 6.4 Warranty tradeoff sensitivity analysis.

sults of a tradeoff sensitivity analysis where the life-cycle cost (LCC) of two candidates is plotted as a function of MTBF.

In the baseline sensitivity analysis, MTBF was clearly the dominant driver; therefore we plotted the LCC of the two candidates as a function of MTBF. Looking at the plot, if the true MTBF is as low as 40 percent of the baseline MTBF, then clearly candidate 1 is preferred since it has a lower LCC than candidate 2 (low is good). At a true MTBF of 50 percent of the baseline MTBF the gap is closing. At a true MTBF of approximately 60 percent of the baseline MTBF the curves cross. From that point on, candidate 2 is preferable. Now it is judgment time. The key question is what will be the most likely true MTBF in the field. The best person to answer that question is the reliability engineer who made the initial estimate of MTBF for the baseline analysis. Based on his rationale for the baseline MTBF and his judgment on how much he could be in error with his original estimates, the program manager or project engineer will make the call. In our example the reliability engineer may admit to feeling confident about the initial MTBF estimate and be highly unlikely to be off by 40 percent on the downside. With that information the decision maker can select the best candidate: in this case, candidate 2.

An additional word regarding the decision process. If the sensitivity analysis procedure described above is not practical because of time or other constraints, care must be taken when choosing the best candidate from the baseline ranking. We need to know not only the ranking but also how close the figures of merit are to each other—the percentage difference. If, for the example given above, the figures of merit for the two candidates are only, say, 2 percent apart in LCC, considering the variability in the overall process, that may well be too close to call on the basis of LCC. In that case, other factors such as reliability or availability may be considered to substantiate the choice. The line between significant and insignificant differences is a judgment call. The tradeoff analysis process is designed as a tool to aid decision-making judgment and not to replace good judgment.

Tradeoff analysis is a powerful tool when applied properly and can be used effectively to trade off the warranty factors to substantiate the optimum warranty statement.

Impact on Warranty Cost

A poorly analyzed warranty strategy could cost a company potential sales or cause the company to incur large remedial costs. If the warranty strategy is not competitive with respect to all the tradeoff factors, sales will be hampered. For example, incomplete or ineffective tradeoff analysis could lead to a recommendation of a 12-month warranty length, where an 18-month warranty would have exceeded the competition and therefore would have been an attractive factor to help increase sales. On the flip side, this same incomplete or ineffective tradeoff analysis could have indicated that an 18-month warranty was the most attractive in terms of sales, but failed to adequately recognize the impact on repair and replacement costs of that extra 6 months of warranty. What we are saying here is, while we need to be competitive with respect to all tradeoff factors, we must also thoroughly investigate the costs of being totally competitive, especially relative to the maintenance and support costs.

The manufacturer must be able to quantify the performance of the product and to understand the product's niche in the marketplace. The tradeoff process and associated analysis provides the tool to thoroughly quantify the performance of the product, and thorough market research helps the manufacturer understand the product's niche in the marketplace. This knowledge is combined with the legal aspects of commercial or consumer warranties to generate the formal warranty statement.

As a final note, lest we forget the buyer, the buyer should study the potential criticality of the warranted equipment to the business and project the cost impact of unscheduled downtime before purchasing the product.

Tradeoff Example

We will illustrate the use of tradeoff analysis to help formulate a warranty statement based on facts. Let's assume that the Gotcha Company produces electronic security systems. They are new in the business.

General Manager I. M. Smart has ordered a market research study of the competition. The results of the study have been presented and are summarized as follows:

- One major competitor: I.C.U. Company
- I.C.U. has been in business for 5 years
- I.C.U. written warranty:
  - 12-month warranty
  - Free replacement policy
  - Covers defects in material and workmanship
  - Replaces defective items
  - On-site service
  - No liability if system shows tampering
  - Arrival on-site within 12 h
  - Restores system within 8 h of arrival

I. M. Smart and her product managers study the I.C.U. Company warranty and, after many hours of deliberation, decide that they, with diligent quality emphasis, can be competitive on all factors. They further decide that since they are new to the business, they need to offer a better warranty than I.C.U. After additional study, it is decided that the only factor on which they feel comfortable they can exceed the I.C.U. warranty is the length of the warranty. It is decided to investigate an 18-month warranty with the same offerings as I.C.U. on all other factors.

I. M. Smart directs her product managers to evaluate the impacts of the 18-month warranty. They decide that since they will offer the same warranty as I.C.U. on all but the warranty length, they need not conduct any tradeoff analyses on the identical factors. Knowing the purpose of the tradeoff analysis as directed by I. M. Smart, they decide to evaluate two candidates: 12-month warranty as offered by I.C.U., and 18-month warranty. They further decide that the basis of comparison should be the levels of performance required while keeping the profit constant. The performance factors evaluated include the following:

- Price of the item
- Market share
- MTBF
- Cost to repair
- Cost of the item

They decide to use a simple model with calculations performed on a calculator. The model includes the following categories:

- Sales revenue
- Repair cost
- Profit

They meet with the warranty team consisting of

- Sales
- Reliability
- Maintainability and logistics
- Manufacturing
- Marketing

to define the required data to input to the model. The following data was identified for both candidates:

- Estimated sales
- Item price
- Estimated usage during warranty
- Estimated field MTBF
- Estimated cost to repair
- Item cost

On receipt and review of the input data, the data was tabulated as in Table 6.1. The five performance factors are analyzed one at a time. The data tabulation shown below is for the case where item price for the 18-month candidate is unknown. The other four data tables would be the same except for different unknowns for the four remaining factors.

Note that, except for items driven by the length of the warranty, the data is the same. The usage figures are based on 4 h per day of operation, every day.

We will show the model calculations for the item price unknown factor and show only the results for the other four factors since they are calculated in a similar manner. Remember, we are keeping the profit constant.

Calculations for the I.C.U. Company are

$$\text{Estimated sales} = 5000 \text{ units} \times \$1000/\text{unit} = \$5{,}000{,}000$$

$$\text{Estimated repairs} = \frac{1460 \text{ h usage}}{2000 \text{ h MTBF}} \times \$200/\text{repair} \times 5000 \text{ units}$$
$$= \$730{,}000$$

$$\text{Estimated profit} = (\$1000 - \$700) \times 5000 \text{ units} - \$730{,}000$$
$$= \$1{,}500{,}000 - \$730{,}000$$
$$= \$770{,}000$$

TABLE 6.1 Tradeoff Example Input Data

	I.C.U. Company	Gotcha Company
Duration, months	12	18
Sales, units	5000	5000
Item price, $	1000	?
Usage, h	1460	2190
MTBF, h	2000	2000
Cost to repair, $	200	200
Item cost, $	700	700

Calculations for the Gotcha Company are

$$\text{Estimated sales} = 5000 \text{ units} \times X = ?$$

$$\text{Estimated repairs} = \frac{2190 \text{ h usage}}{2000 \text{ h MTBF}} \times \$200/\text{repair} \times 5000 \text{ units}$$

$$= \$1{,}095{,}000$$

$$\text{Estimated profit} = 5000X - \$3{,}5000{,}000 - \$1{,}095{,}000 = \$779{,}000$$

$$5000X = \$5{,}365{,}000$$

$$X = \$1073/\text{ unit}$$

The baseline analysis results for all five factors are tabulated in Table 6.2.

Because the team felt very confident with the reasonableness of their input data estimates, the managers decided that the sensitivity analysis was not necessary, and they would make their decision on the basis of the baseline analysis results.

The consensus of the product managers was that if, and only if, the team felt the product could achieve all the necessary levels of the performance factors determined by the analysis, then they would select the 18-month candidate. They questioned each of the team members relative to the results:

- *Sales.* 5000 is projected for Gotcha, but 9506 is achievable.
- *Reliability.* 2000 h MTBF is comfortable; 3000 h is achievable.
- *Maintainability and logistics.* Better equipment access could drive the cost to repair from \$200 to \$133.
- *Manufacturing.* With improved processes, the production cost per unit could drop from \$700 to \$627.
- *Marketing.* I.C.U.'s price is \$1000, but \$1073 is still competitive.

Considering the positive response from all the team members and the fact that the required performance factor values were based on holding

TABLE 6.2 Tradeoff Example Results

Unknown factor	Value for 18-month option
Item price, \$	1073
Marketshare, units	9506
MTBF, h	3000
Cost to repair, \$	133
Item cost, \$	627

the line on profit, they decided to recommend the 18-month warranty length to I. M. Smart. They felt confident in their recommendation on the basis of these considerations and the realization that the analysis they had performed was basically a worst-case analysis.

I. M. Smart was receptive to the tradeoff analysis approach, convinced that the necessary impacts had been investigated. She further agreed with the recommendations and directed Contracts to prepare the warranty statement reflecting the 18-month length and the engineering disciplines, to see to it that the MTBF, unit cost, and repair cost thresholds from the tradeoff analysis are achieved.

It is hoped that the reader noted that the example followed the prescribed tradeoff process as shown in Fig. 6.3. It is further the author's hope that the example was constructive in illustrating how the tradeoff analysis process and the associated results impact the warranty offering.

Chapter

7

Warranty Costing

Introduction

Much discussion thus far has centered around the importance of win-win warranties for both customer and supplier. Going a step further, we have stated that the likelihood of a win-win situation was enhanced by clear, concise, and thorough warranty requirements leading to realistic supplier quotes. In this chapter we will develop the warranty quote process. The techniques discussed are robust relative to any business sector. For purposes of illustration, we will apply the techniques in the government business sector. Consumer and commercial analysis considerations are covered in Chap. 6 since they are generally applied in tradeoffs leading to the development of a formal warranty statement for the specific product.

Warranty Analysis Process

Figure 7.1 shows the warranty analysis process in the government sector relative to weapon system warranties.

In response to the warranty requirement, the government performs a should-cost analysis for the warranty. The should-cost analysis represents what the government feels is a reasonable cost to them to acquire the coverage specified in the warranty requirement. Coincident with the government's should-cost the contractor establishes a cost on what appear to be contractor liabilities in satisfying the warranty requirements. This cost, through judgment, becomes the contractor's price or quote to the government.

The contractor's price is generally a function of the cost to implement the warranty and the risk incurred if the contractor is unable to satisfy all the requirements in the warranty. It should be pointed out that

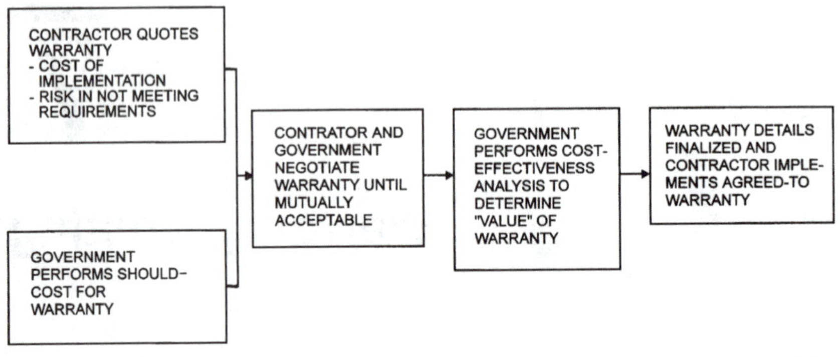

Figure 7.1 Warranty analysis process.

rarely is there agreement between the government's should-cost and the contractor's quote on the first pass. It is likely that the risk factor is the largest single source of disparity between the two numbers. The government, in most cases, will include much less risk money in their estimate than will the contractor. In some cases the government will allow no stipend for risk in their cost estimate. Another source of disparity is the fee factor. The government program people and contracts are instructed in AFR 70-11 not to allow a fee associated with warranties. The two components of cost in the contractor's quote, implementation and risk, will be developed further in a later section of this chapter.

On receipt of the contractor's quote, assuming the quote is unacceptable in its magnitude, the government requests that they and the contractor enter into negotiation on the warranty. In most cases the negotiation will cover several meetings of the two parties and as much as 6 to 9 months of calendar time. First, the requirements are negotiated to assure complete understanding and agreement, then the cost data, including supporting rationale, are negotiated. Eventually, the negotiations produce a mutually acceptable figure. Details of the warranty negotiating process are discussed in Chap. 8.

On completion of the warranty negotiations, the government is required to perform a cost-effectiveness analysis to determine whether the warranty makes good business sense and hence should be implemented. If the analysis clearly shows the warranty to be not cost effective, the government program manager can request a waiver from the Secretary of Defense. If the warranty appears to be cost effective, the said warranty is implemented. Basically, the cost-effectiveness analysis evaluates the projected life-cycle cost (LCC) to the government with and without warranty. The contractor's negotiated warranty quote would cover depot repair costs for the first few years under the warranty. This same cost would represent organic or government depot repair for that same period. This analysis is quite involved and, in many cases, is not performed for that reason. The result is the procurement of warranties with little or no value to the government. This and other issues related to warranty cost effectiveness will be discussed in detail in Chap. 9.

Finally, the warranty administration details are finalized and the contractor implements the agreed-to warranty. Historically, these details are never-ending and are changed frequently through the Warranty Review Board with government and contractor representation. This is a very important phase of the warranty and requires diligent attention and cooperation from both parties. Details of warranty implementation and administration are presented in Chap. 10.

Critical Costing Considerations

Now that we have placed warranty quoting within the framework of the total warranty analysis process, we will turn our attention to the details of the warranty costing process.

Following are critical warranty costing considerations. These items help provide the proper emphasis and establish boundaries to keep the quoting process on track toward the end of a realistic cost, then price, for the warranty.

- Each warranty costing effort is unique.
- Warranty quotes require a concerted, systematic analysis effort.
- A thorough, credible warranty quote must be produced for the customer.
- Quotes based solely on a percentage of production costs should be avoided.
- All warranty quotes should comprehend the negotiating process.
- All warranty quotes should include contractor risk coverage.
- All warranty quotes should have adequate supporting rationale.
- Cost-risk analysis should be performed on all warranty cost estimates.

In my experience, I see numerous occasions where, in lieu of the proper consideration of the uniqueness of a particular warranty or system, the tendency is to force the requirements or system into the mold of a previously quoted warranty. While this may be expedient from a time and labor standpoint, it is not the wise course of action. In my experience, I have never—yes, never—seen any two warranty requirements that were exactly alike. And, I know, there have never been two systems, with their varied operational scenarios and maintenance and support concepts, that were exactly alike. The contractor, in most cases, enters into a high-dollar venture with each warranty. Nothing relative to that particular, unique warranty should be overlooked or generalized to another warranty previously quoted. The uniqueness demands that attention.

Related to the previous factor, each warranty undertaken requires a thorough, step-by-step analysis to preclude omissions or oversimplifications. If there are only 3 days to perform the costing analysis, then as much systematic analysis as can be done in 3 days should be done. The warranty costing exercise provides the constant framework for thorough analysis. Without that framework, key considerations can be overlooked.

It is imperative that the contractor produce a thorough and credible quote to the customer. The subsequent negotiations, and, ultimately the success of the warranty for both sides, goes back to the soundness of the original contractor quote. Many warranties have been penalized from the start owing to the inclusion of excessive cost factors or excessive risk. As mentioned earlier, the success of warranties depends on a partnership, not adversary, relationship between the government and the contractor. The credibility of the quote speaks loudly regarding which relationship is going to prevail in the negotiations, and beyond. Lack of thoroughness, clearly, can backfire on the contractor in terms of inadequate coverage and excessive costs.

The next item is one close to my heart. There seems to be a penchant for "grabbing" published percentage factors and applying them indiscreetly. To illustrate my point, I can recall being present in a program cost review. The program manager was discussing how he arrived at his warranty quote. He, I am proud to say, used our services to conduct a cost-risk warranty analysis to support his quote. In the middle of his discussion, a cohort jumped up and said, "You don't need to go through all that stuff to get the warranty quote." He was immediately challenged by the vice-president for whom the presentation was being made. The vice-president asked him to explain his statement. The explanation included reference to a "canned" percentage of the production costs used for airborne systems. The astute vice-president then asked the gentlemen how many operating hours these airborne systems typically accumulated in the new system's warranty period. The answer, as it turned out, represented only approximately one-tenth of the expected operating hours for the new system. The vice-president then asked the gentlemen what operating hours had to do with production cost. The gentlemen slumped down in his chair as he mumbled "nothing." Had our program manager used this "generic" percentage, he would have grossly underestimated his expected warranty cost. Assuming this percentage to be accurate for airborne systems, this ship system, with its many more hours, hence failures, would not even be in the same ballpark relative to expected repair costs. Needless to say, I was impressed with the vice-president. The bottom line here is that use of canned, generic percentages, with no details, can be very dangerous. Having conducted a thorough warranty analysis, it is permissible to express the quote as a percent of the production cost. The heartburn comes when a bottom-up analysis is not done to substantiate the percentage factor used. These warranties are unique and require unique treatment.

It is important to remember that the initial quote is followed by several negotiating sessions. With this is the realization to the contractor that rarely are quotes negotiated upward. The quote must be set in such a way that gives the contractor some negotiating space.

This advice may sound at odds with the third consideration. In reality it is not. We are not advising contractors to fatten their quotes so as to destroy credibility of the quotes and allow virtually no chance for a cost-effective warranty. What we are saying is, since negotiation is inevitable, contractors must allow themselves the facility to negotiate. They must establish thresholds on certain of the key drivers of their quotes to indicate how far they can negotiate these drivers before their "win" opportunity is jeopardized. It is possible to allow this flexibility in negotiations without injuring the credibility of the quote.

A sister to the previous consideration is discussed next. The flexibility required by the contractor in their initial quote generally takes the form of risk in the quote. As was pointed out in Chap. 2 in the section on risk issues, warranty should involve risks for both the supplier and the customer. The customer should expect some risk money in the quote and challenge the amount in the negotiations. The contractor's quote should have some risk largely because of the timing of the quote relative to the lack of "hard" reliability data on the contractor's system. Challenging the contractor's risk coverage by the customer is not necessarily bad. A contractor who has a defendable position may possibly get to retain a good portion of that risk in the quote. I question the government's tactic of striking all risk from the quote as being unrealistic regarding the contractor's position. A detailed discussion of contractor risk is presented in the next section.

In the interest of speeding and simplifying the negotiating process, the thorough documentation of the supporting rationale for the contractor's quote is a must. Having the supporting rationale in hand can greatly facilitate answering the government's questions. Here again, a plug for the bottom-up warranty analysis. How does one negotiate a percentage of the production cost? Answer, not very well. There is nothing to negotiate. Nothing will close the negotiations faster and send the contractor away faster than having no backup for the contractor quotes. The irony of the situation is that the contractor will eventually end up "backing into" the rationale to support the original quote. Why not do it right the first time? This rationale could include repair labor rates, MTBF allocations, predictions, or analogies to other systems, calculations of the probabilities of passing the performance guarantees related to the EPRs, and repair time estimates.

The final consideration is another one near and dear to me. From my experience, whereas any systematic bottom-up analysis to support the warranty quote is good, a cost-risk systematic bottom-up analysis is even better. The cost-risk analysis allows the contractor to vary key contractor-controllable drivers within reasonable limits to investigate the effects on the bottom-line estimate. In essence the contractor is bounding the bottom line—a sensitivity analysis.

If the contractor has the time and facility, a Monte Carlo simulation approach provides the tool to vary as many parameters as desired at one time—an optimum multivariate sensitivity analysis. This automated technique also allows for a plot of risk as a function of warranty cost, a desirable feature to assist management in setting the warranty price. What we are saying is always perform some kind of sensitivity analysis before setting the warranty price. The warranty quote process is highly estimative and, before setting the price, the contractor needs to know what the driving factors are and how much they drive the bottom line. The Monte Carlo simulation is the ultimate of cost-risk analyses and should be used if at all possible. A detailed discussion of cost-risk analysis, with extensive coverage of the Monte Carlo simulation technique, is presented later in this chapter.

Warranty Cost Components

There are two major warranty cost components: implementation and risk. A warranty quote to the customer is not complete unless there is thorough consideration of both components.

Warranty implementation cost is basically the contractor's cost in carrying out the negotiated warranty. It includes such considerations as fixed costs to establish the contractor depot, unless there is to be an interim contractor support program priced elsewhere in the contractor's total bid package. These fixed costs include test and support equipment, spares, and other related costs to outfit the depot.

The following are typical recurring implementation costs which should be included in the contractor's quote:

- Repair cost
- Transportation cost
- Administration cost
- Field service cost

The repair cost includes all labor and material to accomplish repair of the warranted item to restore it to a serviceable condition. These costs include failure verification, active repair, repair verification, and testing to produce a fully qualified spare unit, and all material used to accomplish the repair. Typically, repair cost is the most significant cost component of the total warranty cost.

Transportation cost is that cost to ship the warranted item to the contractor and back again to the customer's place of use. Depending on how the warranty requirement is worded, the contractor may be responsible for one-way or round-trip shipping costs. While there is no standard approach, many of the requirements state that the government will ship the failed item to the contractor and the contractor is responsible for shipment of the repaired or replaced item back to the government's place of designation, usually the place of use. The transportation cost is typically not a driver in the total warranty cost. If, however, the maintenance concept has typically large and heavy units such as line-replaceable units (LRUs) or weapon-replaceable assemblies (WRAs) coming back to the contractor's depot for repair under the warranty, the transportation cost can become much more significant.

Administration cost includes all costs for repair documentation, processing of warranty data, preparation of required status reports and data items, monitoring of warranty status, and participation on the Warranty Review Board (WRB). In many cases administration cost is underbid, especially if the contractor is bidding on a warranty for the first time. There are many "hidden" costs within administration, and the contractor must develop an administration flowchart to avoid omitting any costs. The government has found, from their side, that historically they have underestimated administration costs when projecting their portion of the total warranty cost.

There is great concern among contractors when bidding a warranty about how their warranted equipment will be handled by service maintenance personnel before it is shipped to the contractor. Despite stated exclusions regarding the improper handling of warranted items by service personnel in the requirement, contractors feel they need first-hand surveillance of the handling of their equipment. Obviously, improper handling which is not visible to the contractor from external inspections can result in an increased number of returns from the field for repair. If equipment surveillance can be accomplished by one of the contractor's people already in the field, the contractor should not quote field service costs. If, however, there is no current coverage by the contractor's people at particular sites where the new warranted equipment is to be used, the contractor will generally include some field service labor in the quote. Field services cost is generally not a warranty cost driver, unless there are several field service people to be assigned, and/or the rate of pay for these people is excessively high.

While implementation costs are concrete and are therefore more easily explained, equally important in the warranty quote are the costs associated with risk to the contractor in offering the warranty. Let us state up front that the inclusion of risk money in the quote is not an indication that the contractor cannot or will not meet the warranty requirements. The contractor fully intends to meet the warranty requirements but, since the equipment is not yet built or tested, the contractor takes a risk in saying, at the time of the quote, that the requirements will be met. It is important that we understand the difference before we proceed.

Generally, the contractor's risk is included in two cost categories of the quote:

- Redesign/retrofit
- EPR (performance guarantee) penalties

It should be pointed out that risk dollars can also be included in the implementation cost. For example, the contractor's estimate of the expected number of returns for repair cost may be based on a MTBF value lower than the requirement if the contractor's reliability estimates indicate difficulty in ultimately meeting that requirement. This risk money embedded in the repair cost should be uncovered by the customer during negotiations and thoroughly discussed as to the degree of its legitimacy.

Redesign/retrofit remedies are generally related to the failure of the contractor to meet one or more of the EPR requirements. If this is the case, then it is legitimate for the contractor to include redesign/retrofit costs in the quote. Knowing how often the EPRs are measured, estimating the probability of not passing the EPR requirement, and estimating the cost of a redesign/retrofit action, the contractor can estimate the risk cost relative to the redesign/retrofit remedy. If, however, redesign/retrofit remedies are not tied to failing the EPR requirements, the customer may challenge their inclusion in the contractor's quote.

Many contractors do not include a cost for redesign/retrofit in their quotes because they feel it is an admission that they cannot design and build what the customer wants. If redesign/retrofit costs are included in the quote, they should have adequate supporting rationale to facilitate the negotiation of such costs. As a footnote, where redesign/retrofit costs are included in the contractor's quote and are not related to EPR requirements, it is very difficult for contractors to decide on the magnitude of these costs since they have no indication of how often they will not be in compliance on the EPRs. Without this kind of rationale, contractors will have difficulty negotiating this cost with customers. The bottom line is that when redesign/retrofit remedies are tied to EPR requirements, they are legitimate risks and can therefore be costed as risks. Without this relationship, contractors are forced into a much more tenuous position. Finally, there are cases where redesign/retrofit remedies are tied to lot inspection reject levels, and can therefore be included in the quote.

The other cost-risk component in the quote is that of EPR penalties implemented as performance guarantees. EPR compliance, as seen in Chap. 5, may have penalties other than redesign/retrofit remedies. These penalties are generally tied to spares assets or dollars.

The final example discussed in Chap. 5 is an excellent example of the use of free spares from the contractor as a penalty for EPR noncompliance. In this case, the contractor was held to a criterion involving both MTBF and TAT. If the combined effect of MTBF and TAT performance during a measurement period was below the requirement measured in spares, and if, currently needed spares quantities exceeded the current existing stockage, the contractor was required to provide free spares equal to the difference of the two quantities. Let's assume that the contractor was faced with this requirement. How would the contractor assess the cost to include in the quote to cover this risk?

Fortunately, the supplier requirement offered was quite explicit as to the penalties and under what conditions they would be liable for the penalties. It remains for the contractor to decide how difficult it will be to meet the MTBF and TAT requirements. Contractors accomplish this by putting values they feel comfortable in meeting in the given equations and calculating the results. Obviously, there are many possible combinations of MTBF and TAT where the contractor would be liable for free spares and a normal "what if" analysis could not begin to evaluate them. Ideally, a Monte Carlo cost-risk analysis where many random draws from simple distributions for MTBF and TAT could be accomplished and the results calculated for, say, 1000 runs, would be desirable. This technique provides an estimate of the probability of incurring the situation where the contractor would be liable, and calculates the penalty spares quantities, as applicable.

From this example it is easy to see the importance of a thorough requirement from the customer. Without this detailed rationale, contractors would have no way of estimating their risk in meeting the EPR requirements and passing the performance guarantee. In assessing contractor risk in meeting the EPR requirements, there must be the capability of estimating the probability of passing or not passing the requirements. Without this statistical approach tied to a well-defined set of criteria, the contractor's chances of successful risk coverage during negotiations are severely hampered.

Before we leave the risk component of warranty cost, it is necessary to mention incentives. If the EPRs contain incentives for the contractor, they may play in the probabilities of exceeding the requirements and therefore reflect profit in their quote (negative costs), or they can set the negative cost equal to zero (no penalty). This is up to the contractor's discretion.

Benefits of Bottom-up Costing

At the risk of being repetitious, we need to drive home the benefits of a detailed warranty costing analysis. Referring again to the last example in Chap. 5, a less-than-detailed analysis, perhaps a percentage approach, would not have given the necessary visibility into the contractor's risk in not passing the EPR requirements or relative to the redesign/retrofit remedy. There is simply no way that this risk and associated cost could be properly assessed without a detailed analysis. Needless to say, these penalties involve significant possible expenditures for the contractor and deserve detailed attention.

In highlighting the potential costs associated with risk issues we are not minimizing the importance of bottom-up analysis relative to the implementation costs. They, too, deserve a detailed evaluation, especially repair costs, so that the entire quote is done in as thorough a manner as is feasible.

If you have already decided that this is a sensitive area with the author, you are right on. I have heard many a lament from contractor program management regarding "why didn't we do it right the first time" or "we left money on the table." So, do the warranty costing analysis right the first time—do a bottom-up analysis!

Warranty Quotation Process

For purposes of this discussion, we will assume that the contractor has thoroughly reviewed the warranty statement from the government and understands the requirements. Additionally, we will assume that the contracts group has been contacted to see if similar warranties have been quoted. Figure 7.2 shows the warranty quotation process.

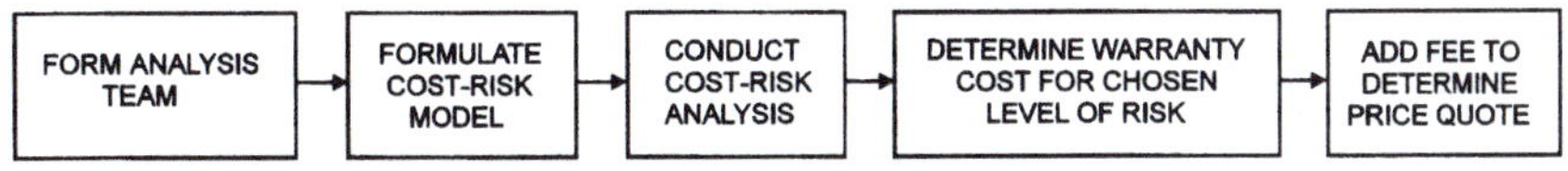

Figure 7.2 Warranty quotation process.

Form analysis team

The initial step is to form the warranty analysis team. Figure 7.3 illustrates the warranty analysis team.

Generally, the team's function is to help interpret the warranty statement, provide data for the cost-risk analysis, conduct the analysis, and set the warranty quote. A detailed description of their responsibilities is provided in Chap. 11. For purposes of this chapter, except for the cost analyst and program manager, the most important function of the team is providing input data for the cost-risk analysis.

Steps 2 through 4 will be considered the warranty costing process and step 5, the pricing.

Warranty costing process

In formulating the warranty cost-risk model and performing the related analysis, several different approaches can be taken:

- *Sensitivity analysis.* After establishing a baseline warranty cost with nominal single point estimates, parameters can be varied to determine whether they are strong drivers (univariate and multivariate).
- *Bounding technique.* Estimate the limits associated with key cost drivers and evaluate the cost model at these extremes to give the minimum and maximum warranty costs. The final warranty cost will be a value within these lower and upper limits.
- *Beta distribution approach.* Three different costs are estimated for each cost category: a most likely cost, lowest cost, and highest cost.

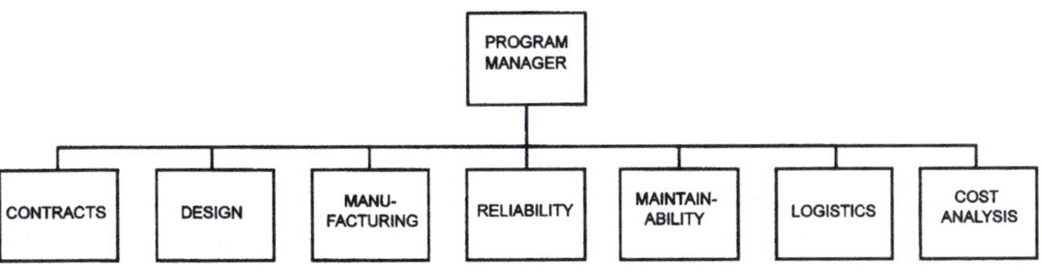

Figure 7.3 Warranty team.

This allows the determination of confidence limits and standard deviation of the total cost.

- *Monte Carlo simulation technique.* Parameters known to be drivers and are probabilistic can be described in the cost model using statistical distributions. This method provides an expected value as well as the relationship between risk and warranty cost.

The choice of the approach may depend on time availability and the reception of management to the technique. From the author's experience, the Monte Carlo simulation technique has proved to be fairly easy to implement and well received by management. It can be programmed in Lotus on a PC and can be linked to external statistical software for random-number generation, histograms, and other statistical outputs. Additionally, the technique is robust to the timing of the analysis—a very attractive feature. Regardless of when in the life cycle of the equipment a warranty cost-risk analysis is performed, the Monte Carlo technique is applicable. The only factor that changes over time is the range of the Monte Carlo variables. The closer to fielding of the warranted equipment, the tighter the estimated range of the variables since better estimates of the data can be made. With the highly estimative nature and the related variability of the data associated with the warranty quote process, the Monte Carlo technique provides an excellent tool for the purpose of warranty cost-risk analysis. It is a realistic and informative approach.

Because of the overall benefits of the Monte Carlo technique, it is the author's judgment that steps 2, 3, and 4 of the warranty quotation process can best be developed within the framework of a detailed discussion of the technique. One further word before we discuss the Monte Carlo simulation technique. This technique is not intended to replace but to supplement management judgment in the warranty pricing process. Managers can visually trade off warranty cost and risk to maintain a balance between competitive pressures and acceptable risk.

Model description. The model to be described has been used to establish the risk associated with different warranty pricing policies. The equipment supplier's cost evaluation must include all cost elements that are potentially the supplier's responsibility in order to assess overall liability. This would include any penalty costs or possible incentive fees that are unique to the supplier's warranty requirement. Table 7.1 lists the cost elements considered in the model with brief definitions. These are cost, not price.

A key ingredient in evaluating the warrant cost is an estimate of the

TABLE 7.1 Cost Element Definitions

Element	Definition
Repair cost	All contractor costs associated with item repair including direct labor, direct material, support labor, and overhead
Shipping cost	The cost of transporting the item to and/or from the contractor's facility to a customer supply point
Administrative cost	The cost of administering the warranty including periodic warranty meetings and reviews, settling warranty issues, tracking and monitoring the progress and status of warranted items, and any other administrative tasks associated with the warranty
Field service cost	The cost of providing field personnel to assure government compliance with field repair of the item, assist in the repair activity as required, and serve as the interface between the supplier and the customer
Redesign/retrofit (R/R) cost	All contractor costs for modifying an item because of its nonconformance to a warranty performance requirement such as mean time between failures (MTBF), repair turnaround time, or retest OK rate; this may include development cost, manufacturing cost, and cost for retrofitting all or part of the existing inventory
Penalty cost	The cost of additional resources incurred by the supplier for not complying with specific performance requirements such as mean time between removals (MTBR) or logistics downtime
Incentives	This is the "reward" for exceeding performance requirements; the value of this incentive award, monetary or otherwise, is subtracted from the total warranty cost

expected number of item returns that occur during the warranty period. This is based on the anticipated item field MTBF and the operating scenario. The form of the repair cost algorithm depends on the warranty repair clause. For example, the cost to repair all failures must be included if the supplier is to repair all returned units (except for specified exclusions) under a fixed-price up-front agreement. However, if the supplier is responsible only for failures over an allowable level based on a comparison of actual MTBF and specified MTBF, then only this group should be included in the repair cost. In some warranties the contractor's liability does not begin until enough operating hours have been accumulated to ensure that the data is statistically significant. For some types of systems, such as missiles and munitions, the operating MTBF may not be a critical performance parameter. Instead, the storage MTBF will drive the repair cost.

A repair cost algorithm applicable to avionics systems in which the contractor is liable for all repair costs during the warranty period is shown in Eq. (7.1).

$$\text{RC} = (\text{FLS}) \cdot [(\text{URC}) + (\text{RTOK}) \cdot (\text{UROC})] \tag{7.1}$$

where RC = total repair cost for an item
FLS = expected number of true failures of item
URC = average cost to repair an item (unit) failure
RTOK = retest OK rate stated as a fraction of true failures
UROC = average cost for processing a unit that retests OK

$$\text{FLS} = \frac{(\text{HRS})(N)(\text{NI})(K)(\text{MON})}{\text{MTBF}} \tag{7.2}$$

where HRS = operating hours per system per month
N = number of systems under warranty
NI = quantity of specific item per system
K = factor used to degrade the predicted MTBF based on field experience
MON = warranty duration in months
MTBF = predicted operating MTBF for item

Equation (7.2) is evaluated for each item in the system and summed to arrive at the total system repair cost.

Shipping cost, which considers the one-way cost to transport an item from the supplier facility to the customer, can be described as

$$\text{SC} = (\text{FLS})(1 + \text{RTOK})(\text{WT})(\text{SR}) \tag{7.3}$$

where SC = shipping cost for item
FLS = expected number of true failures of item
RTOK = retest OK rate stated as a fraction of true failures
WT = weight of item in pounds
SR = cost to ship an item in dollars per pound ($/lb)

If the contractor is responsible for the round-trip cost, then a factor of 2 should be included. Again, each item is evaluated separately to arrive at the total system transportation cost.

The warranty administration cost and field service cost are calculated in a similar fashion, as shown in Eqs. (7.4) and (7.5).

$$\text{AC} = (\text{ACP})(\text{NAP})(\text{WAP}) \tag{7.4}$$

where AC = total warranty administration cost for system
ACP = average cost per month per administrative person
NAP = number of administrative worker-months per month
WAP = length of warranty administration period in months

$$\text{FS} = (\text{FCP})(\text{NFP})(\text{WFP}) \tag{7.5}$$

where FS = total field service cost for system
FCP = average cost per month per field service person
NFP = number of field service worker-months per month
WFP = length of warranty field service period in months

Any supplier who is noncompliant with respect to performance requirements or systemic coverage may be required to take corrective action to improve the system by incorporating a no-cost redesign-retrofit remedy (R/R). The number of R/Rs required for a system and the cost per R/R can be highly variable. The R/R cost must consider both potential hardware and software modifications and include the costs for redesign, production, and possibly retrofitting the existing inventory. The R/R cost equation is

$$R/R = (NRR)(CRR) \tag{7.6}$$

where R/R = total redesign retrofit cost
NRR = expected number of R/Rs
CRR = average cost per R/R

The final cost elements addressed by the model are the penalty costs and incentives unique to each warranty contract. They are usually tied to essential performance parameters of the system such as MTBF, repair turnaround time (TAT), repair time, retest OK rate, or equipment availability. The typical form for the penalty-incentives equation is to compare the resource requirement using the measured value for the performance parameter versus the contractual value. If the performance requirement is not satisfied, then additional resources (such as spares, support equipment, or penalty dollars) are provided by the supplier to compensate for not satisfying the warranty provision. A supplier whose performance exceeds the requirements will receive appropriate remuneration. A sample spares penalty equation based on a MTBR performance requirement could take the following form:

$$SP = (ASP - CSP)(UC) \tag{7.7}$$

where SP = cost of penalty spares for item
ASP = required spares based on actual MTBR and specified TAT
CSP = required spares based on specified MTBR and specified TAT
UC = item unit cost

A point estimate of the warranty cost can be obtained using the described model by substituting the expected value for each input parameter and summing the cost elements. However, since there is a high degree of uncertainty associated with the key drivers of the warranty price (e.g., MTBF, RTAT, and repair cost), a point estimate provides no information regarding the risk associated with the estimate.

The Monte Carlo method requires that the key inputs controllable by the supplier be assigned a probability distribution that characterizes the expected variability in the parameter. Then, random values from these distributions are selected and used in the cost model to arrive at a warranty cost. This process is repeated often enough to obtain a frequency history of the warranty cost. A cumulative probability distribution can then be derived that will show the probability of exceeding any selected warranty cost (risk).

Figure 7.4 shows the Monte Carlo process. Although the method is simple in concept and readily understandable by management, it does require additional information to be secured from individuals providing the input data. Instead of supplying only a most likely value for a parameter, they must provide additional data to define the parameter probability distribution. Selection of the appropriate distribution can be based on empirical data or one of the many available theoretical distributions. Experience has shown that the simple uniform and triangular distributions will satisfy the majority of cases for describing input parameter uncertainty.

Any dependencies between random variables must be considered when using the Monte Carlo method. If all variables are totally independent, then a separate random number is used to obtain a sample

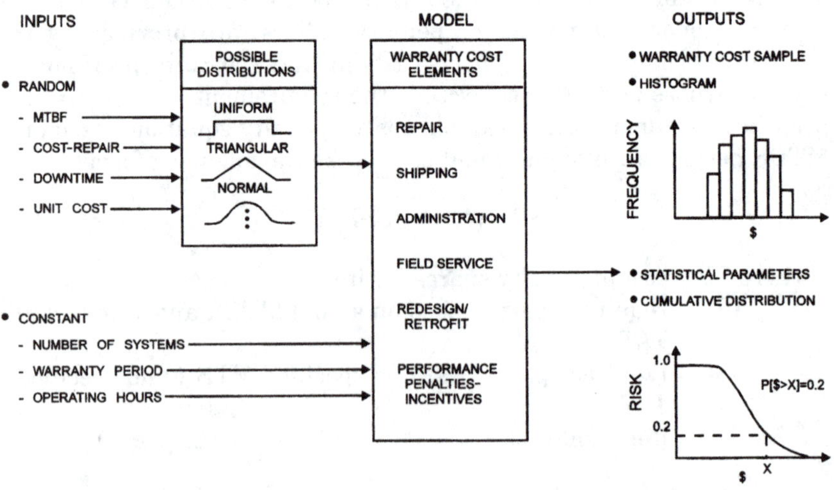

Figure 7.4 Monte Carlo warranty process.

value from each parameter distribution. However, if dependency exists between two variables and their interrelationship is not considered then the results of the Monte Carlo simulation will be unrealistic.

Techniques exist for measuring the goodness of fit between the simulation results and a theoretical distribution such as the normal distribution. If the comparison is successful, then the statistical properties of the theoretical distribution can be used to characterize the simulation results (i.e., tolerance limits). This is especially convenient for the normal distribution whose properties are common knowledge. The absence of normality in the output results does not invalidate the methodology. Figure 7.5 summarizes the Monte Carlo statistical steps.

The results of the Monte Carlo analysis can be used to ascertain the probability of exceeding a selected warranty cost (see Fig. 7.5). The relationship between the selected warranty cost and the baseline warranty estimate, derived from using nominal values, establishes the management reserve that is set aside to cover risk. The resulting proposed warranty price must be weighed against competitive pressures and/or customer monetary limitations to achieve the proper balance between supplier risk and a winning proposal.

Model implementation. After warranty requirements have been defined and the relevant cost elements determined, the task of translating these into cost-generating equations specific to the computer used begins. The Lotus 1-2-3 spreadsheet software is selected because of its availability and widespread usage in the PC arena. The add-in software called @RISK is installed in Lotus 1-2-3 to provide Monte Carlo simulation capability.

The variables to be randomized and the distribution(s) to be used are selected on the basis of information received from the persons estimating the input data. @RISK allows the use of 24 different probability distributions. These distributions can be added to any number of cells and formulas throughout the spreadsheet, and can include arguments that are cell references and expressions. The uniform distribution is used frequently owing to its conservative nature and because it requires only minimum and maximum values to be estimated. All values within the range specified have equal likelihood of being selected dur-

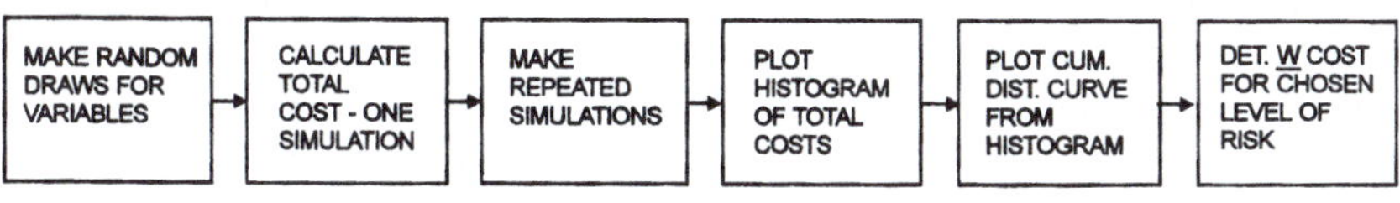

Figure 7.5 Monte Carlo statistical steps.

ing random draws. If it is known that more values will occur about a certain value within the range, then the triangular distribution is used. Interdependencies are modeled by the @DEP and @INDEP functions, which identify the sampling pair and provide either a positive or negative correlation.

Once the model is ready for simulation, @RISK is accessed and the @RISK menu appears at the top of the spreadsheet. The items on the menu are arranged in logical order. After selecting the sampling mode under the SETTINGS heading, the model outputs are designated. The ITERATIONS option allows the number of iterations of the model equations to be specified. Time permitting, the more iterations performed, the greater the confidence in the process stability.

The simulation is begun using the EXECUTE option. With each iteration, @RISK draws a new sample from each distribution function in the worksheet and recalculates the entire spreadsheet using the new values. If the UPDATE-DISPLAY option is selected during setup, each worksheet recalculation will show the results from that iteration. This can be a time-saving feature since a programming error can be spotted and corrected without waiting for the whole set of iterations to be completed. It is especially valuable when the worksheet is large, and a set of iterations may take as long as 2 h to complete. If no error is made, calculations can resume using the CONTINUE function under the EXECUTE heading.

The results from each iteration are captured to create the output probability distributions. These distributions can be viewed using the RESULTS function. Selecting CURRENT under that heading calls up a histogram of the first output specified. The next histogram will appear if NEXT is chosen and the cumulative curve associated with any histogram can be viewed by selecting TYPE. The statistical information of each of these distributions as well as the graphs can be saved for future reference. Or, if a printout of a graph is desired immediately, the PRINTGRAPH feature under the REPORT heading can be used. This feature eliminates the need to return to the original spreadsheet and then out of 1-2-3 to get to the LOTUS PRINTGRAPH mode.

The expected warranty cost is shown at the top of each distribution chart. This is equivalent to approximately 50 percent probability on the cumulative curve. An excellent feature called @RISK SETTING TARGET VALUES is provided when viewing the cumulative chart. It gives the exact warranty cost for a desired risk (or probability that the warranty cost will exceed the cost quoted). This feature can also give the risk for a specified warranty cost driven by the competition or by the customer's budget. Thus, different risks and warranty costs can be investigated to arrive at a final competitive quote. It must be emphasized that a supplier's judgment is still vital when this methodology is used. This tool allows the supplier to logically quantify the warranty cost and, perhaps, couple that with other information not modeled in the technique to produce the quote. An example of how this is accomplished is shown in the following section.

Example requirements. The following example is taken from a typical warranty requirement for multilot avionics equipment with a stepladder delivery schedule and a warranty period that is truncated after 6 years.

The RFP states that all equipment supplied must be free from defects in material and workmanship. The supplier must comply with essential performance parameters such as MTBR and depot dock-to-dock TAT. The seller must repair all returns. If a systemic defect occurs, the supplier must repair, redesign, and retrofit the equipment at no increase in cost to the buyer. The supplier is also responsible for all shipping costs from the supplier's facility to the buyer. If the average actual MTBR does not meet the average guaranteed MTBR, then additional spares must be provided based on the difference between the initial spares (computed using the guaranteed MTBR) and the actual number of spares needed to satisfy the average actual MTBR. For every 48 h that the average actual TAT exceeds the average guaranteed TAT, the supplier must pay a penalty of $500. Both performance parameters will be monitored every 3 months during the warranty period.

Example assumptions. From the warranty requirements stated above, the supplier must cover all repair costs, regardless of whether the MTBR requirements are met. Thus, the repair cost shown in the table of warranty costs (Table 7.2) covers labor and material as well as repair cost risk, where repair cost risk is the risk coverage for the repair

TABLE 7.2 Warranty Costs

Cost category	Cost, $	Percent
Repair cost	1,200,000	20
Shipping cost	150,000	2
Administration	375,000	6
R/R cost	1,350,000	22
Field service	225,000	4
MTBR penalty cost	1,875,000	30
TAT penalty cost	975,000	16
Total warranty cost	6,150,000	

of additional failures caused by not meeting the MTBR. The supplier pays for transportation of the repaired item back to the buyer and also incurs cost in administration and field service. According to equipment performance, hardware and software redesign and retrofit may be needed. The cost for this is covered in the R/R cost. Coverage for the two different penalties is given by the MTBR and TAT penalty costs.

Example analysis. Lotus 1-2-3 and @RISK equations were formulated from the warranty requirements and @RISK was set to perform 1000 iterations. Each iteration draws random values for MTBR, unit cost, cost per R/R, TAT, and several other variables on the basis of their respective distributions. A set of costs is calculated from these values and constitutes one pass of simulation. Table 7.2 shows a sample simulation pass.

The sum of these costs is the total warranty cost for that iteration and this is chosen as the output of interest. Table 7.2 shows that the repair cost makes up 20 percent of the total warranty cost. However, this is only one simulation pass. All cost categories may vary as percentages of the total warranty cost with each pass. Each pass is captured to provide the final histogram and cumulative cost-risk curve of the total warranty cost, as shown in Figs. 7.6 and 7.7.

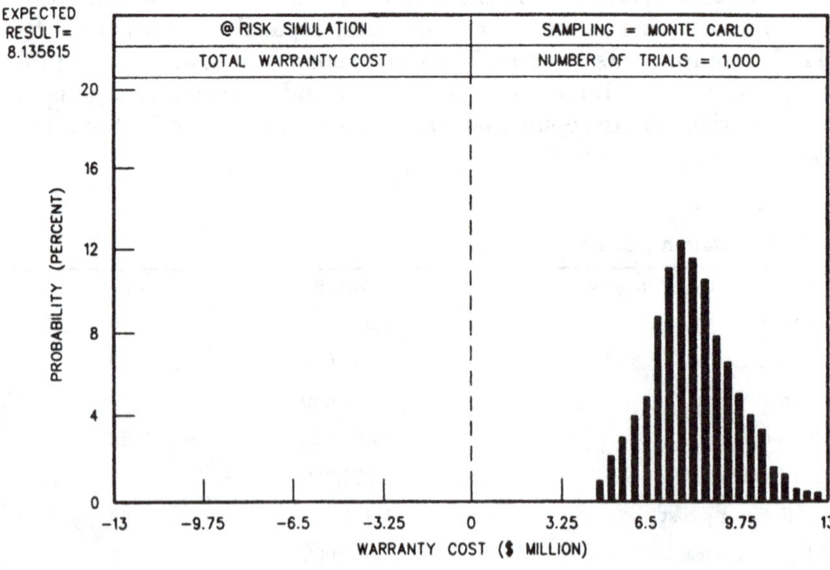

Figure 7.6 Histogram of total warranty cost.

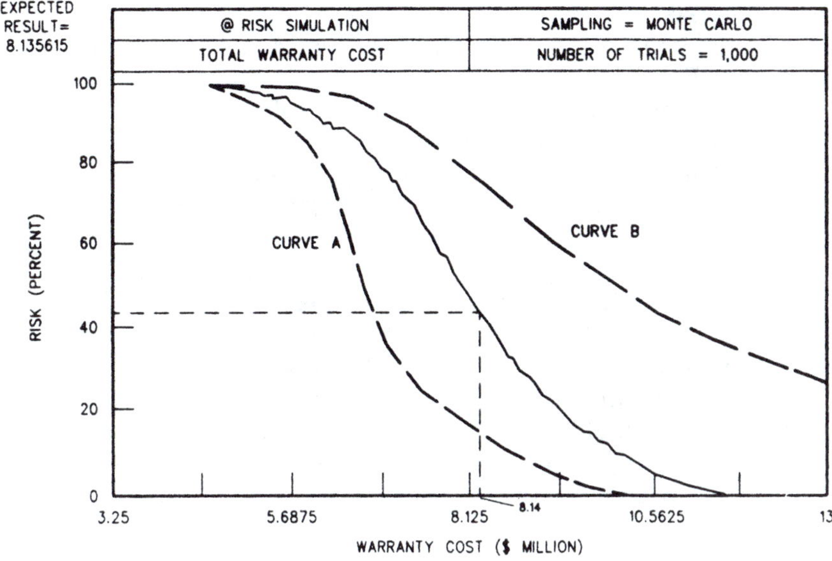

Figure 7.7 Warranty cost-risk curve.

Example conclusions. At the expected total warranty cost of $8.14 million, given on the top left-hand corner of the figures, there is approximately a 50 percent chance that the warranty may cost the supplier more. The total warranty cost ranges from about $4.47 million at 100 percent risk to $11.78 million, at zero risk (see Fig. 7.7).

The supplier may decide to use 20 percent as an entry point on the *y* axis of the graph to obtain the total warranty cost. By tracing that across and down to the *x* axis of the graph, the supplier will arrive at a value of $9.34 million. This method of determining the corresponding cost for a risk can be fraught with errors and is cumbersome because of the awkward scale on the *x* axis. A more accurate value can be obtained when using the @RISK SETTING TARGET VALUES feature.

However, the $9.3 million value may seem high, so the supplier tentatively sets the warranty quote at $9 million to stay competitive. To know what kind of risk is being taken by quoting the warranty at that value, the supplier uses the @RISK SETTING TARGET VALUES feature and enters 9.0. The risk is 24.6 percent. If this risk is acceptable, the $9 million value is used to make the quote. If not, the supplier may select a series of different risks to see which is most acceptable.

The warranty cost-risk curve of Fig. 7.7 has a nominal slope. Therefore, the supplier sees a moderate amount of change in cost with each percent of risk difference. A steeper slope indicates that a large change in risk results in only a small difference in total warranty cost. (See curve *A* in Fig. 7.7.) If this is the result despite large variances specified in the input data, then the supplier is able to assume a very low risk without tremendously increasing the warranty cost. A more gradual slope has a large change in cost for every percent difference in risk, as illustrated by curve *B* in Fig. 7.7. The tighter the range of the input distributions, the steeper the slope of the cost-risk curve.

Besides observing how the warranty cost will vary when different risk values were chosen, the supplier may reexamine the input data to determine how to refine them in light of what is known about the cost of the warranty. The supplier may modify assumptions or data ranges for cost elements that drive the warranty cost.

The warranty analysis is conducted in this manner until the supplier is satisfied that adjusting the data and varying the risks have provided a reasonable warranty cost.

Lessons learned. Through the development and subsequent use of the subject model, certain key findings have emerged. Managers find the graphical display useful in relating warranty cost and risk. As the program progresses, the limits of the input data defining the appropriate distributions should get tighter, and/or more exact distributions could be used to improve the warranty cost estimates. Experience indicates that the uniform and triangular distributions serve well, particularly early in the system program.

The team approach to warranty analysis is the most cost-effective approach. Technical specialists in their respective disciplines assigned to a particular program are the best source of input data. No single person or group can effectively price a warranty.

If the program management does not understand the basic concept of simulation and the need for considering the variability of warranty estimates, they will favor a mean value, point estimate approach. This approach may not be ideal but it is superior to the "percent of production cost" strategy.

Finally, this cost-effective analysis technique is not a substitute for management judgment. The technique is, however, a valuable tool to aid the decision maker in assessing the risk associated with various pricing strategies or the cost associated with a chosen level of risk.

Future developments. While the technique has proved practical and beneficial, much remains to be done in promoting wider acceptance and use in the management community. This acceptance will be enhanced by gaining a better understanding relative to interpretation of results—gleaning all possible lessons from the data, including the impact of parameter dependencies.

Establishment of warranty price (quote)

Once the Monte Carlo cost-risk analysis on contractor costs has been performed, all that remains is setting the warranty price for quotation to the customer. Contractor management must decide whether a fee is to be added to the cost, and if so, how much. The pressures of competition weigh heavily on this decision. They may decide that the warranty cost is conservative for their chosen level of risk and that a fee is not necessary. Having experience in prior negotiations, they may decide to include a reasonable fee and hope that they can salvage at least some of it in negotiations. As pointed out earlier in this chapter, government program managers are directed per AFR 70-11 not to accept a fee in the warranty quote, and it may be a useless exercise to include a fee.

At best, this is a judgment call. Armed with the cost-risk analysis results—the associated risk for a chosen warranty cost or a warranty cost associated with a chosen level of risk—the contractor can, however, make a much more informed decision as to the amount of fee appropriate to that warranty situation and having made that decision, can now set the warranty price for quotation to the customer.

Cautions

The warranty quotation process is labor-intensive and often requires estimates for parameters that have not been well defined. It is therefore essential to carefully evaluate all the assumptions made to ensure that they are reasonable and workable. This can be accomplished only if the engineering cost analyst works together with the project team members (who are the estimators) and relies on them to resolve issues encompassed in their area of specialty.

An important consideration when performing a warranty simulation is dependencies between random variables. Not accounting for this correlation can lead to understatements of the variance in the cost-risk curve.

When no data is available and the data estimator team has little confidence in their estimates, it is better for the seller to use more conservative estimates (take less risk), provided their quote remains competitive. This is especially true for warranties that have performance requirements which are stringent and where the seller may be severely penalized for nonconformance.

Since the warranty may be critical to contract award, it is often beneficial to commence warranty analysis as soon as the decision is made to pursue a contract. Otherwise, the warranty team may not be able to formulate a saleable warranty that demonstrates strong commitment to quality at a competitive price.

Since most warranties have different implementation requirements and penalties or incentives, it is important for the engineering cost analyst to conduct a comprehensive warranty analysis. Only then should the selected warranty cost be compared to the warranty cost based on more general techniques such as a percentage of the total production cost. In this way, the percentage serves only as a guide and the analyst has allowed the unique requirements of that warranty to drive the cost.

To illustrate the dangers of the indiscriminate use of generic percentages in lieu of a unique comprehensive warranty analysis, Tables 7.3 through 7.5 present warranty summaries for Army, Navy, and Air Force systems, respectively.

From Tables 7.3 through 7.5 it is important to note the difference among the three systems. Relative to remedies for correcting defects, the Navy and Air Force system warranties indicate the contractor's responsibility for all failures (failure-free warranty), while the Army system warranty is a threshold warranty, with the contractor's liability beginning on the 3184th failure. Also the Army warranty limits the contractor's liability to $21 million, while the Navy and Air Force warranties have no liability cap.

The Navy warranty requires the contractor to bear all transportation costs, while the Army warranty requires the government to pay for the transportation costs. The Navy warranty indicates a redesign remedy, while the other two do not.

With respect to the basic warranty language, the Navy requirement stresses engineering change proposals (ECPs) as the means of correcting deficiencies, otherwise they are similar. Considering essential performance guarantees, the Army and Navy warranties cover operational capabilities and some mention of reliability, while the Air Force warranty is clearly an MTBF guarantee. Finally, these systems are quite different in mission and complexity.

Now let's focus on the background section of each warranty summary, specifically the warranty price and price of warranted items (total production cost).

Suppose we are seeking a "generic" statistic to represent the warranty price in lieu of a unique bottom-up analysis. Further, let us suppose that the decision was made to use published percentages of warranty price as a percentage of the total production cost. Calculating these percentages for the three systems yields 0.04, 2, and 17.4 percent, respectively. Without a thorough study of the differences cited above, including the fact that all had different durations, we are left with a problem. Which percentage should I use? There is such a wide disparity among the three numbers that any choice of a percentage could represent a sizable price difference in the warranty quote from that obtained in a detailed bottom-up analysis.

The fact is that many times these percentages are used without a

TABLE 7.3 Army Apache AH-64A Helicopter

ARMY APACHE AH-64A HELICOPTER

Background

Procurement Organization:	U.S. Army Aviation Command
Contract Date:	9 April 1985
Price of Warranted Items:	$666,358,898
Warranty Price:	No cost except administration ($274,000 FY 1985)
Production Phase:	FSP - 4th year
Warranty Period:	2 years or 240 flight hours, whichever occurs first, for material and workmanship, essential performance, and design and manufacturing

Remedies for Correcting Defects

Contractor repairs or replaces failed depot components of 138 aircraft after the 3,183rd allowable failure up to liability cap of $21M. Contractor reimburses the Government for repair or replacement of any parts due to defects in materiel and workmanship that occur on a lot basis. Contractor reimburses the Government for the cost of repair and replacement if the contractor fails to repair or replace promptly. Contractor does not bear transportation costs.

Basic Warranty Language

Coverage: Notwithstanding inspection and acceptance by the Government of supplies furnished under this contract or any provision of this contract concerning the conclusiveness thereof, the Contractor warrants, for the period set forth in para c, that any aircraft, procured under this contract, including all warranted components and lot defects on non-depot repairable parts installed on such aircraft:

(1) Will meet performance requirements specified in this Warranty Clause.
(2) Will be free from all defects in material and workmanship at the time of delivery that would cause the warranted items to fail to meet any performance requirements specified in this Warranty Clause.
(3) Will conform to the design and manufacturing requirements set forth in Section C.1 of this contract, consistent with the contractor's approved Quality Assurance System.

Liability: The contractor shall be liable for all failures and direct and resultant damage caused thereby, not excluded from coverage, to the extent set out in this clause and not otherwise limited herein or elsewhere in this contract. The contractor's obligation under this clause shall be to repair or to absorb the cost of repair of failed warranted components beginning with the 3,184 repair. Contractor's maximum liability shall not exceed $21,000,000. Included within this limited liability is a separate $1,000,000 limitation on resultant damages as defined herein.

Essential Performance Guarantee

Requirements:	Contained in technical manuals for operation and maintenance with failure rates no greater than allowed by the AH-64A system specification MTBF. Evidence of failures of depot-repairable assemblies must not exceed 3,183 failures from the 138 warranted aircraft. All parameters are related to field performance checks such as rate of climb, gauge readings, or satisfactory maintenance tests.
Validation Means:	Operation and maintenance checks.

study of the backup information. In our example we have three military systems. Which percentage should I use to represent the present military system? Or, having come across only one of these percentages, should I use it? In either case, I could be off by as much as three orders of magnitude from the "true" percentage. Is that adequate accuracy? I think not!

The bottom line is that there is no substitute for an analysis where the unique requirements of that warranty on that system drive the warranty cost, leading to a reasonable warranty quote to the customer. Beware of published percentages when quoting a warranty!

TABLE 7.4 Navy Mine Neutralization System

NAVY MINE NEUTRALIZATION SYSTEM

<u>Background</u>

Procurement Organization:	Naval Sea Systems Command
Contract Date:	July 1984
Price of Warranted Items:	$24,909,272
Warranty Price:	$498,186
Production Phase:	Initial
Warranty Period:	3 years for material and workmanship and for design and manufacturing/performance

<u>Remedies for Correcting Defects</u>

Contractor repairs or replaces defective parts. Contractor corrects defects by redesign. Contractor reimburses the Government for the cost of repair and parts replacement if the contractor fails to repair or replace promptly. Contractor bears transportation costs.

<u>Basic Warranty Language</u>

Notwithstanding inspection and acceptance, the Contractor guarantees that:

(a) Specified components are designed and manufactured to conform to the performance requirements described in the weapon system specification.

(b) Specified components, at the time of acceptance, are free from defects in material and workmanship which would cause components to fail to conform to the performance requirements of this contract.

Notwithstanding any provision of the contract, the Contractor is responsible for preparing Engineering Change Proposals (ECPs) and for all aspects of implementing ECPs required to correct deficiencies.

<u>Essential Performance Guarantee</u>

Requirements:	Weapon specification examples include depth, neutralization rate, detection range, and reliability.
Validation Means:	Specifications, first article test, factory acceptance test, environmental stress tests, Sea Board trial, test and monitoring of system prior to and after acceptance, but prior to use.

TABLE 7.5 Air Force ARN-118(V) TACAN

AIR FORCE ARN-118(V) TACAN

Background

Procurement Organization:	Air Force, Electronic Systems Division
Contract Date:	July 1975
Price of Warranted Items:	$72,023,206
Warranty Price:	$12,506,985
Production Phase:	Initial
Warranty Period:	4 years (RIW and MTBF guarantee)

Remedies for Correcting Defects

Under RIW, contractor repairs or replaces every covered failure. Under MTBF guarantee, contractor determines causes of nonconforming MTBF, develops and implements corrective action, and provides consignment spares in the interim.

Basic Warranty Language

Under RIW, the system will be free from defects in design, material, and workmanship, and will operate in its intended environment in accordance with contractual specifications and for the warranty period set forth in the contract. Under MTBF Guarantee, the system will achieve a MTBF value equal to or greater than the following: 500 hours (1 through 12 months), 625 hours (13 through 24 months), and 800 hours (25 through 48 months).

Essential Performance Guarantee

Requirements:	MTBF.
Validation:	Operate time is measured by elapsed-time indicators, and failures are those covered under the RIW.
Results:	Final results show that system MTBF exceeded 1,000 hours, well above the highest guarantee value. Warranty administration worked well, and the warranty program is considered a model RIW program.

Chapter

8

Warranty Negotiating

General Considerations

Effective negotiation is a crucial step in the realization of mutually beneficial warranties. While legislation provides the basic structure for warranty applications, it by no means assures success of warranty programs. This chapter describes the warranty negotiating process and emphasizes the respective responsibilities of the supplier and customer to help ensure cost-effective warranties. As discussed earlier, in both the commercial and government acquisition processes, negotiations are generally involved. In the commercial sector, the results of the negotiations become part of the purchase agreement. In the government sector, on military programs, the results of the negotiations become part of the contract. Since the supplier and customer dynamics relative to negotiating would not be expected to be significantly different between the commercial and government sectors, the military warranty negotiating process is described as representative. We will use the terms *contractor* in place of *supplier* and *government* in place of *customer* throughout the discussion.

Negotiation Process

Negotiation follows the evaluation by the government of the contractor's warranty quote. The government evaluates the contractor's quote against their should-cost analysis in the period between receipt of the quote in the engineering and manufacturing development (EMD) phase of the program and approximately one year before the completion of the EMD phase. The actual negotiations must occur prior to low-rate initial production and deployment, since the law applies to production units. Negotiation helps to ensure the contracting of cost-effective warranties by involving give-and-take between the government and the winning contractor.

Typically, there is difficulty on the contractor's part in interpreting the warranty requirement, and the contractor's quote could reflect that uncertainty in terms of an inflated dollar amount, despite the risk in so doing of losing the contract. The negotiation affords the opportunity for the government to challenge the quote, leading to a mutually agreeable dollar amount. This final dollar amount is then input to the government's cost-effectiveness analysis.

Figure 8.1 depicts the basic military warranty negotiating process. In Fig. 8.1, the customer is the government. The process begins with the warranty requirements developed by the government during demonstration-validation appearing in the EMD request for proposal (RFP). The clearer the requirement, the higher the likelihood of a cost-effective warranty for the government. Many warranty negotiations are needlessly delayed or deterred because of vague warranty requirements. Simplicity of requirements should be the goal. Since warranty requirements definition by the government is a continuous task across several program phases, it is shown in Fig. 8.1 to be concurrent with the interpretation of the requirements by the contractor.

The contractor interprets the warranty requirements and prepares the associated warranty quote for production units for inclusion in the

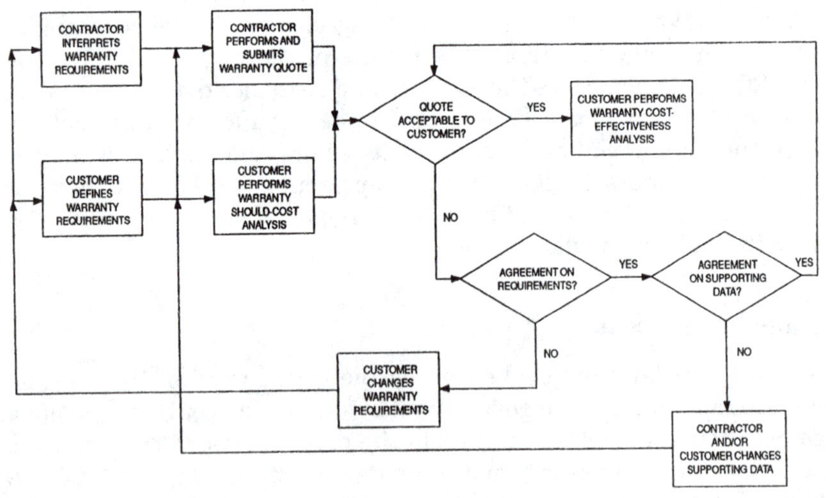

Figure 8.1 Warranty negotiating process.

EMD proposal. The quote includes the cost of implementation of the warranty at the contractor's facility and the risk cost to the contractor in offering a fixed-price warranty on equipment yet to be built or tested. An important factor in the give and take of warranty negotiation is the understanding that warranties involve risk sharing. Neither government nor contractor should be expected to bear all the risk; warranty is risk sharing.

It should be pointed out that, in some cases, government program managers are instructed to contract for no-cost warranties. There is no such thing as a free warranty, as can be clearly demonstrated in the commercial sector. Since the point in time when the warranty requirement is set, the government should be performing a warranty should-cost analysis to provide a yardstick with which to judge the reasonableness of the contractor's quote.

The contractor's warranty quote is received by the government, so noted, and analyzed. Depending on the evaluation criteria, the warranty price could be a significant factor in the award of the EMD phase contract. The winning contractor executes the EMD and subsequent phases knowing that the stated warranty will be executed on the initial and full-rate production units, on deployment.

At a predetermined point in the execution of the EMD phase contract, and prior to production, the negotiating activity begins. Referring to Fig. 8.1, it is possible that the initial quote from the contractor is acceptable to the government, but this is not likely, particularly if the warranty price was not directly involved in the LCC evaluation process in contract award. The government and contractor negotiators meet to reach agreement. It should be pointed out that the negotiating process can be quite lengthy and require numerous meetings. Patience is the byword. An important consideration that can shorten the duration is that of having good substantiation of both the quote and the should-cost. Both sides must have done their homework.

Contractors should not expect the negotiating process to be facilitated when their quote is based solely on a percentage of the total production cost. There must be bottom-up substantiation of that percentage in order to allow effective negotiation.

As the negotiations begin, an atmosphere of partnership, not animosity, must be established. This can most easily be done if both sides realize that they can "win" if the warranty is properly negotiated. The government can procure products with improved performance and reduced support costs, and the contractor can achieve a profit from the warranty, as well as other less-tangible customer satisfaction benefits.

Per Fig. 8.1, if the initial quote is not acceptable to the government, the requirements are negotiated. Discussion of the realism of supporting data is moot if there is not mutual understanding and agreement on the requirements. If there is an impasse on the requirements, a reevaluation and restating of the requirements may be necessary, resulting in a requote from the contractor and another look at the should-cost by the government. Requirement issues could include the duration of the warranty, conditions and/or remedies, definition of a defect or failure, and the frequency and process for essential performance parameter measurement. When agreement is reached on the requirements, attention shifts to the supporting data for the price quote. Again, the absence or paucity of supporting data makes the reasonableness of the contractor's quote or the government's should-cost extremely difficult to judge. The discussion of supporting rationale can be quite lengthy and require many recesses and return trips. Several requotes could result from discussion of or agreement on supporting data. Examples of this rationale include assumptions regarding the MTBF that can be achieved over the warranty period, costs per repair at the contractor's facility, and costs per redesign or retrofit when a defect is uncovered. At some point in the warranty negotiations, agreement will be reached on the requirements and the supporting data and, hence, per Fig. 8.1, the quote is acceptable to the government.

At the conclusion of the negotiation, the government is required to perform an analysis to determine the cost-effectiveness of the warranty using the negotiated dollar value and reliability and readiness indices. In essence, the government evaluates the life-cycle cost (LCC) and system effectiveness with and without the warranty to make this determination. If the results conclusively indicate that the warranty would not be cost-effective to the government, the government program manager can request a waiver from the Secretary of Defense. History has shown that, in many cases, the cost-effectiveness analysis was not performed and the warranty implemented with no "value" to the government. In these cases, the negotiations could have been quite thorough and effective, but negated because of the lack of a cost-effectiveness analysis. Both effective negotiation and thorough cost-effectiveness analysis are necessary for value-added warranty contracting.

Application

The negotiating process described herein applies to equipment termed weapon system by United States Code, Sec. 2403 of Title 10, and further defined in Defense Federal Acquisition Regulation (DFAR) Subpart 46.770. Exclusions under the weapon system definition include support equipment, training devices, ammunition, government-furnished equipment (GFE), and certain commercial equipment.

Cautions

The final negotiated requirements should meet the requirements of the law. Revised requirements must adhere to the spirit and letter of the law, as best as can be determined. The final negotiated dollar figure should be the result of negotiating in good faith on both sides. Without honest give-and-take, the win-win goal for contractor and government with warranties will not be achieved. Everything is negotiable within the law. The reaching of a mutually agreeable and mutually beneficial dollar figure is likely to be a very arduous task and can be greatly facilitated by the existence of detailed supporting rationale on both sides. This support data includes MTBF estimates, repair cost estimates, asset penalty-incentive estimates related to the EPRs, and administrative labor cost estimates. Neither side can expect thorough or expedited negotiations without this level of detail. As a final note of caution, per Chester Karrass, in negotiations, you do not get what you deserve, you get what you negotiate.

Importance of Negotiations

As discussed in Chap. 3, the warranty statute is subject to numerous interpretations. Even the most meticulously developed warrant clauses could evoke misunderstanding between the government and the contractor. Typically, there is very little communication between government and contractor relative to clarification of the requirements during the prequote period. The combination of these two limitations, in virtually every case, produces an impasse between what is intended by the government and what is supplied by the contractor. Negotiation is the necessary process to bridge this impasse.

Negotiation affords the opportunity for both parties to discuss the warranty requirements and supporting data for the price quote in a structured face-to-face situation. In this situation, there is adequate time for the necessary give-and-take that must take place before a mutually agreeable warranty price can be settled. It is important to note that negotiation typically does not produce a result that both parties wanted, but rather a result that both parties can accept. Without the face-to-face give-and-take afforded by the negotiating process, it is virtually impossible to achieve a win-win result relative to the total warranty process.

Because the contractor is in business to make a profit, and because the government is attempting to get the most value from the taxpayers' dollars, a price disparity is almost certain. Negotiation focuses on the disparity and illuminates the details of the disparity to facilitate agreement in the most effective manner—face-to-face communication.

Example

Situation

In this example, the customer is a prime contractor and the supplier is a subcontractor. Per previous discussions, the prime contractor passes the requirements on to the subcontractor. The prime contractor for the Blue Two aircraft has requested a quote from the radar subcontractor per the following warranty requirement.

1. *Conditions* The subcontractor guarantees that (*a*) for a period of three (3) years from acceptance by the prime contractor each item (radar) will conform to the specified design and manufacturing requirements as delineated in the production contract and essential performance requirements as delineated in paragraph 2(*b*) below, and (*b*) for a period of three (3) years from acceptance, each item (radar) at the time of delivery is free from all defects in material and workmanship which would cause the item to fail to conform to the specified performance requirements of this production contract and all other subsequent modifications thereof entered into by the contracting officer and the contractor.
2. *Remedies*
   (*a*) *Repair or replacement of failed items.* In the event of a failure of the contracted item to meet the conditions specified in 1(*a*) and 1(*b*) above, the subcontractor shall effect such action as to correct the failure at no additional cost to the prime contractor. In addition, the subcontractor shall prepare and furnish to the prime contractor data and reports applicable to any such corrective action.
   (*b*) MTBF Guarantee. Under normal field conditions, the subcontractor's warranted item shall achieve the field MTBF requirement of 215 h as calculated by dividing total operating hours accumulated by the total number of failures experienced at the conclusion of the warranty period. Failure to achieve the required MTBF shall result in the subcontractor providing, at no additional cost to the prime contractor, one spare unit (20 percent of radar unit cost) for every 5 h of MTBF below the field MTBF requirement.
3. *Exclusions* Failures resulting from the following causes are excluded from the warranty:
   (*a*) Combat damage
   (*b*) Damage from fire, flood, crash, explosion, or an act of nature, unless the subcontractor's negligence contributed thereto
   (*c*) Environments more severe than those delineated in the equipment specification

(*d*) Unauthorized repair attempts by the prime contractor after acceptance
(*e*) Operating hours in excess of those negotiated

4. *Transportation.* The prime contractor will bear all the costs of shipping the failed items to the subcontractor, and the shipment of the items back to the using organizations.

The prime contractor has performed a warranty should-cost analysis per the requirement and the following conditions and assumptions.

- 45 operating hours per system per month
- 250 installed systems
- 1884 allowed failures (total over warranty period)
- Expected field MTBF = 215 h
- Expected unit cost = $300,000
- Expected labor (worker-hours) including repair and documentation per failure = 28 worker-hours
- Labor rate per worker-hour = $50
- Material cost per failure = $900
- 3 ECPs at $450,000 each
- Equivalent of 2.5 persons per year for administration at $40,000 per year
- Equivalent of 1.5 persons per year for field service at $42,000 per year
- Total warranty program duration 108 months
- Subcontractor assumed to pass MTBF guarantee test—no penalty

The radar subcontractor has performed a warranty analysis and arrives at a quote price per the requirement and the following conditions and assumptions:

- 45 operating hours per system per month
- 250 installed systems
- 1884 allowed failures (total over warranty period)
- Expected field MTBF = 160 h
- Expected unit cost = $300,000
- Expected labor (worker-hours) including repair and documentation per failure = 35 worker-hours
- Labor rate per worker-hour = $55

- Material cost per failure = $1075
- 2 ECPs at $325,000 each
- Equivalent of 3 persons per year for administration at $42,000 per year
- Equivalent of 2 persons per year for field service at $48,000 per year
- Total warranty program duration 108 months

There is a wide disparity between the prime contractor's should-cost and the radar subcontractor's quote. The result is that negotiations are in order.

Negotiation

According to Fig. 8.1, the warranty requirements receive first consideration. Using criteria established in Chap. 3, the requirement is incomplete. According to the requirement, it would appear that the prime contractor has specified a failure-free repair warranty. Only through phone conversations does the subcontractor learn that the prime contractor really wants a threshold warranty based on a field MTBF of 215 h, and prices the warranty accordingly.

There is also confusion as to when the 215-h MTBF must be achieved. Because of these two significant ambiguities and its incompleteness relative to other warranty elements, the requirements are judged to be less than adequate. According to Fig. 8.1, the customer should clarify these omissions and ambiguities in a revised requirement for new should-cost and new quote evaluations.

Assuming that the requirements have been made clear and complete, an agreement on the requirements has been reached. Interest now focuses on the supporting data for the should-cost and quote. The prime contractor and subcontractor agree on the operational data gained through review of the specifications. Using the operating hour data and the required MTBF of 215 h, the threshold number of failures is determined to be 1884. From this point on, there are differences in the supporting data between the prime contractor and the subcontractor.

Because the MTBF of 215 h must be achieved at the end of the warranty program under the MTBF guarantee and the subcontractor wishes to include risk money to cover the uncertainty of achieving 215 h, the subcontractor bases the repair costs on 160 h, an average over the entire warranty period. The repair cost for the prime contractor is based on a 215-h MTBF.

Both the prime contractor and the subcontractor assume administration and field service costs, but using different numbers of people and pay rates. The repair worker-hour estimates and the labor rates are also not consistent. Also, the same holds for the repair material cost. Both sides assume the need for redesign and retrofit through ECPs to achieve the required MTBF, but the number and cost per ECP are not consistent.

Relative to EPR penalties, the prime contractor assumes that the subcontractor will meet the MTBF requirement, and hence no risk cost is included in the should-cost for this category. The subcontractor, on the other hand, includes risk in this category by using a very conservative 160-h MTBF. The subcontractor views the MTBF penalty as being potentially severe and wishes to be well covered.

As mentioned in Chap. 7, all the appropriate costs—repair, transportation, administration, field service, redesign and/or retrofit (ECP), and EPR penalty—have been accounted for in both the should-cost and the quote. The subcontractor transportation cost is zero per the requirement.

It is likely that if this negotiation were to take place, much of the time and attention would focus on the MTBF value relative both to resulting repair costs and EPR penalties. MTBF drives the warranty cost most heavily in this example and, in fact, in many negotiations.

According to Fig. 8.1, the negotiations continue until agreement has been reached on the supporting data. Once agreement on the requirements and the supporting data for the price is reached, the price is mutually acceptable and the price negotiation is complete. Having a final price, the government then must perform the final cost-effectiveness analysis to determine warranty value. Details on warranty implementation and administration are yet to be negotiated.

The reader is encouraged to assume they are the contract negotiator for the prime contractor and formulate a series of questions to determine the cost categories addressed in the subcontractor's quote, the dollar value of each category, and the supporting data for each dollar value. Also, how would you determine what MTBF the subcontractor actually used in the analysis, why, and the resultant impact on risk cost? (You would likely point out to the subcontractor that even though the MTBF value is calculated at the end of the warranty period, the data used cover the entire period, and therefore 215 h is an average MTBF over the entire period, not 160 h.)

The reader is then encouraged to assume that they are the contract negotiator for the subcontractor and formulate an explanation for the use of the 160-h MTBF rather than the required 215 h as the basis for repair and EPR penalty costs. (While the subcontractor may have legitimately reasoned that 160 h was the average MTBF over the entire warranty period, having been challenged by the prime, they agree that 215 h is the correct average. Their argument for use of 160 h MTBF for repair and EPR penalty costs would reduce them to an ultraconservative position relative to risk.) Also, provide a defense of the repair worker-hour estimates and labor rates and the administration and field service people levels and yearly pay rates.

Negotiation can be an arduous and lengthy process, but a necessary process to the achievement of a win-win warranty result. Clear and concise requirements, adequate supporting rationale, and a spirit of cooperation, however, can ease the burden of the negotiation process. The byword for both supplier and customer is—"Be prepared!"

Chapter

9

Warranty Cost Effectiveness

General Considerations

In Chap. 1 we briefly discussed cost effectiveness as a figure of merit for the value of a warranty. In this chapter we will develop and apply the concept.

While cost-effectiveness analysis techniques can be applied in any business sector, our discussion will be relative to the government military sector. Owing to unique requirements, the military sector affords an excellent vehicle in which to develop the subject. We will use the terms *cost benefit* and *cost effectiveness* interchangeably.

Assessing the Value of Warranties

We use the term *cost effectiveness* in our everyday activities. We say things like "that was a cost-effective choice" or "we need to be cost effective in our decisions." Without being aware of it, we apply the concept of cost effectiveness often in our dealings. Let's look more closely at cost effectiveness.

- What? Measure of benefits received versus expenditures incurred.
- How? Integrate reliability, maintainability, supportability, availability, and life-cycle cost (LCC).
- Why? We need a figure of merit for warranties.

As the integration of reliability, maintainability, supportability, availability, and LCC, cost effectiveness provides a good measure of competing warranties. What we are really saying is that cost effectiveness is a valid tool for determining warranty value.

As discussed later, the tool will be applied in the context of helping to determine whether a particular warranty is cost effective to acquire as compared to not acquiring the warranty. Without some assessment of both the cost and the benefits of the warranty, the decision to acquire or not acquire a warranty is incomplete. The benefits portion is often difficult to quantify, but necessary for a valid decision. It is important to point out that in some instances relative to military warranties, a LCC analysis is sufficient rationale on which to judge the cost effectiveness of a proposed warranty.

The armed services have been instructed to obtain warranties only when they are of value. The Army has further said that they will only acquire warranties that meet the requirements of the law and make good business sense—that is, have value. Cost-effectiveness analysis is the chosen tool with which to determine warranty value for the customer.

Requirements

As presented in DFARS Subsec. 246.770-7, it is DOD policy to obtain only cost-effective warranties under the warranty law (Sec. 2403 of the Title 10 U.S. Code). If a specific warranty is not considered cost-effective by the contracting officer, a waiver request should be initiated following procedures described under DFARS Subsec. 246.770-8. In assessing the cost effectiveness of a proposed warranty, an analysis must be performed to consider both the qualitative and quantitative costs and benefits of the warranty. Costs include warranty acquisition, administration, enforcement, user costs, weapons system LCCs with and without a warranty, and any costs that result from limitations imposed by warranty provisions. Costs incurred during development specifically to reduce warranty production risks should also be considered. The cost-benefit analysis (CBA) must also consider logistical and operational benefits expected as a result of the warranty as well as the impact of any additional contractor motivation provided by the warranty. Where possible, comparisons may be made with the historical costs to obtain and enforce similar warranties on similar systems. The analysis should be documented in the contract file.

The preceding requirements from DFARS Subsec. 246.770-7 are exhaustive in scope. There is, however, no direction on the details of the analyses. To respond to the DFARS requirements, each service established guidance on the conduct of warranty CBAs. For example, the Air Force published instructions for weapon system warranty CBAs in AFR 70-11. AFR 70-11 provides instructions to Air Force program managers relative to the planning and conduct of the warranty program.

The document states that a CBA must be done to determine whether the contemplated weapon system warranty (WSW), which will be in the production contract, is cost effective. A CBA must be done, even though the contractor may propose a "no-cost" WSW, to compare the government's cost of administering and enforcing the WSW to the potential benefits to be derived from the proposed WSW. AFR 173-15, paragraph 4-7, provides Air Force guidance for conducting the CBA, as well as when the Air Force should accomplish and update the CBA.

Several DOD organizations have reviewed WSW cost-effectiveness analysis (CEA) history since the requirement became effective in 1985. The findings have shown that either CEAs are not being performed or CEAs are deficient in detail. The result in either case is the lack of substantiation for the warranty purchase. Some reports have stated that the government is spending millions of dollars on warranties every year when waivers should have been requested for many of them. Very few waivers have been requested since the inception of the public law. Some have estimated less than 3 percent of the warranties have requested waivers. The law of averages would say that figure is far below what would be expected. Lack of valid rationale for the CEAs seems to be the major reason why many military organizations have not requested waivers.

During a 1989 General Accounting Office (GAO) assessment of the DOD's administration of its warranty programs, it concluded that the military services had not yet established fully effective warranty administration systems and recommended that the Secretary of Defense's office extend its oversight of warranties. The GAO went on to say that as a result of the lack of fully effective warranty administration systems, the DOD has little assurance that warranty benefits are being fully realized.

Of specific interest, they stated that waivers of warranty law requirements were generally not being sought by the procurement activities included in the GAO's review. They further stated that, despite the existence of sound cost-effectiveness models, problems were being experienced in performing CEAs; thus, the activities are not in a position to know whether they should seek waivers.

To put the problem in proper detailed perspective, one of the four principal findings was that procuring activities included in the GAO review either have not been performing CEAs or have prepared analyses that do not adequately support conclusions that proposed warranties are cost effective. As a result, procurement activities were not considering waiver requests in their decisions on proposed warranties because their analyses did not provide a convincing basis to support requests for waivers in cases where warranties may not be justified because they would not be cost effective.

As a result of these findings, the GAO recommended that the services devote more attention to performing appropriate CEAs to be able to determine whether waivers should be requested.

It is clear from the above findings that CEA is an important process and this process is either not being performed or being performed inadequately by the services.

Cost-Effectiveness Analysis Concept

Purpose of CEA

From prior discussion, it is the intent of federal legislation that the services acquire only cost-effective warranties. CEA is the tool for evaluating the comparative cost effectiveness of warranty and no-warranty candidates to assist in the selection of the candidate with the most value to the government. CEA should be a function of LCC and system effectiveness (SE) to provide the most thorough rationale to support the warranty, no-warranty decision. As stated earlier, in some cases it is feasible to evaluate only the comparative LCC of the candidates. In this case the decision is made solely on the basis of cost. If qualitative information exists on the comparative reliability or availability, then it should bear on the decision, along with LCC. Ideally, the support analysis should be totally quantitative to provide the maximum decision support. In any case, there must be sufficient and valid rationale to support the acquisition of the warranty or the waiver request.

Timing of the CEA

The program manager must ensure that the warranty CEA is initiated when the technical design is established sufficiently to allow LCC estimation. The CEA may be accomplished as early as during the demonstration and validation phase, and updated during EMD and source selection or negotiations for the production contract that contains the weapon system warranty. It is preferable to begin the CEA process as early as possible. As a very minimum, the CEA should be accomplished before release of the request for proposal (RFP) for the production contract that will contain warranty provisions and updated after receipt of proposals to reflect the contractor's proposed price. The timeframes are illustrated in Fig. 9-1.

In Fig. 9.1, MS I through MS IV are the major program milestones. At each milestone the program undergoes a detailed status review within the DOD and must get DOD approval before proceeding to the next program phase. During the demonstration and validation phase a determination is made relative to the need for a warranty based on the criteria of the law. If a warranty is needed, preliminary CEA effort is begun. In the engineering–manufacturing development (EMD) phase

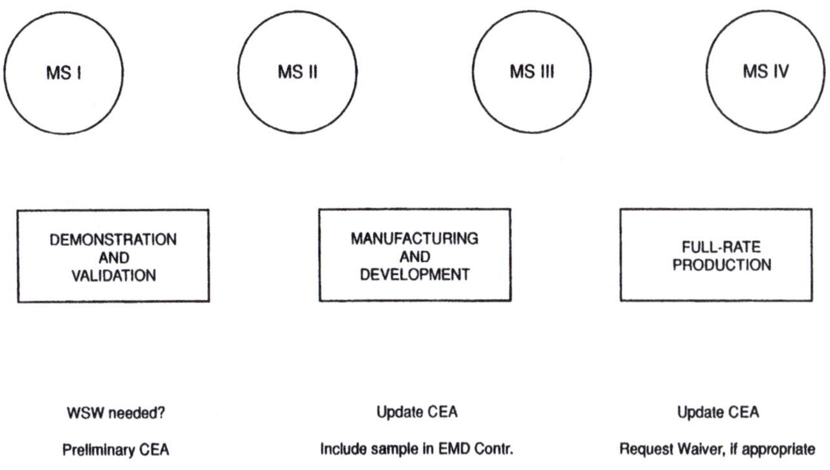

Figure 9.1 Cost-effectivess analysis in system life cycle.

RFP a sample warranty is included, in accordance with which the contractor submits a not-to-exceed quote which is input to the updated CEA. Prior to the production-deployment phase, the contractor's quote is finalized and the warranty price is negotiated. The final price is input to the final CEA, and the decision on warranty acquisition or waiver is made prior to system production and deployment.

Application of LCC

The cost of the warranty impacts the acquisition portion of LCC. It is part of the weapon system cost and increases total acquisition cost. The potential benefits derived from the CEA impact the operation-support (OS) portion of LCC. The benefits can be broken down into the applicable categories such as labor, material, support equipment, and workforce. Benefits computed from the CEA could reduce these OS categories.

Team concept

The team concept is an important factor during the CEA process. The warranty team should be formed early in the program to ensure cohesion in the warranty CEA process. Coordination and mutual understanding of technical terms, timeframes for coverage, and funding considerations are important. The various team members may use warranty terminology differently. A simple example is calculations made on a calendar-year basis rather than a fiscal-year basis. The use for the information obtained should be made clear by the analyst performing the CEA.

Data for evaluation

The data used for evaluation comes from many sources within the program office and, depending on the phase of the program, the data may be very limited. Reliability and maintainability engineers can provide projections, cost projections can be made, and potential benefits can be estimated. If a similar system exists, the historical data can be used to make estimates. Until the design becomes more firm, only preliminary estimates can be made. Figure 9.2 presents the typical CEA data flow among team members.

As depicted in Fig. 9.2, the CEA analyst is the hub of the data flow. The analyst defines the required data for the model to each of the program functions. On receipt of the input data, the analyst performs the CEA and interprets the CEA results to aid the program in their acquisition or waiver decision. A more thorough discussion of the customer warranty team and the associated data estimation and flow is provided in the CEA process section of this chapter.

Performing the CEA

The specific methodology for performing a CEA will vary depending on the type of warranty selected, the type of weapon system, the terms and conditions exercised by the contracting officer, the essential performance characteristics of a weapon system, and the identification and measurability of various types of costs. Since each warranty is different, the methodology is tailored to meet the specific requirements.

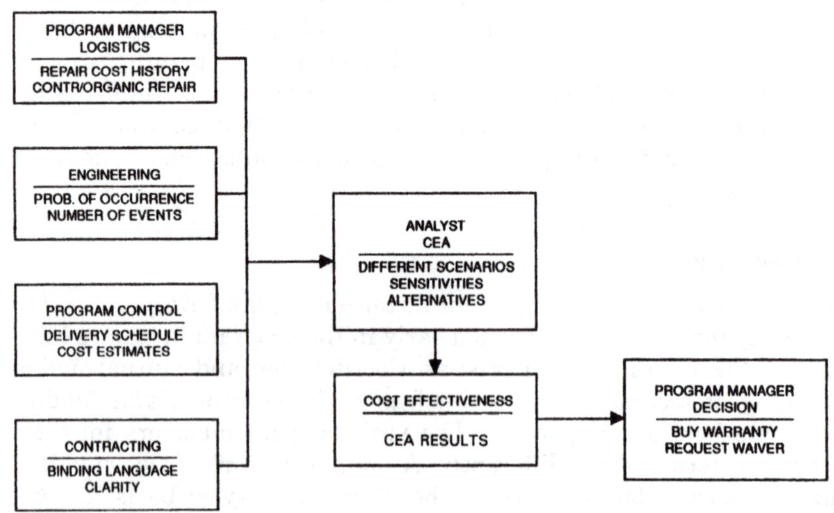

Figure 9.2 Cost-effectiveness analysis data flow.

A LCC model should be used as the basis for determining areas impacted by the purchase of a warranty. A separate model, for warranty cost only, can be developed using a spreadsheet. Sensitivities should be performed using technical assessments from engineers. For example, a certain MTBF or shop visit rate is projected, and sensitivities can be performed to vary the number on the basis of a range of values. Risk assessment is also documented and can be a subjective evaluation of high, medium, or low technological risk. There are some warranty cost-effectiveness models available, including ARINC's Warranty Decision Support System and the U.S. Army's Warranty Model (WARM). Often a model must be developed or adapted for the specific application.

As suggested by DFARS Subsec. 246.770-7, the expected benefits from the warranty should be compared with warranty acquisition and administration costs. The analysis should examine the expected costs with the warranty versus the cost expected if the weapon system were supported under normal organic (service) support conditions or possibly contractor support services. If the program manager or contracting officer does not consider a specific warranty proposition cost effective based on the CEA, a waiver request should be initiated under DFARS Subsection 246.770-8.

Per DFARS Subsec. 246.770-7, the warranty CEA is required to be documented and made a part of the contract file. Documentation should identify data sources and explicitly present the methodology and approach used to estimate costs and benefits over the life of the weapon system. Documentation should be sufficiently complete so that another analyst could reperform the procedure and reproduce the same results.

Warranty benefits

Warranty benefits encompass more than the "snapshot in time" during the warranty coverage period. The tangible benefits derived during the coverage period equate to the contractor-repaired items. However, the intangible benefits must also be considered. These intangible benefits include better reliability and maintainability and increased availability. Intangible benefits are difficult to convert into dollars. However, if the warranty CEA process begins early in the program, these improvements should be documented as part of the analysis.

Analysis framework

In this section the basic framework for a warranty CEA is presented. First we will present the LCC framework, followed by system effectiveness, and finally the cost-effectiveness framework.

LCC. The time value of money (i.e., discounted or inflated money) is not considered in this discussion for the sake of simplicity. The discussion assumes constant dollars over the life cycle. In Eq. (9.1), LCC is the total cost to acquire and sustain a system over its lifetime (this includes the costs of development, investment, operation, support, and retirement) and WCB is the warranty cost benefit.

$$\text{WCB} = \text{LCC}_{\text{NW}} - \text{LCC}_{W} \tag{9.1}$$

where LCC_{NW} is the LCC with no warranty and LCC_{W} is the LCC with warranty.

If LCC is the only decision metric (effectiveness not considered), then WCB must be positive or zero for the warranty to be cost effective. Separating LCC into two levels, one can establish a basis for evaluating warranty price:

$$\text{LCC}_{W} = \text{WP} + \text{LCC}_{\text{WP}} \tag{9.2}$$

where WP is the warranty price and LCC_{WP} is the LCC to the customer exclusive of warranty price combining Eqs. (9.1) and (9.2):

$$\text{WCB} = \text{LCC}_{\text{NW}} - (\text{WP} + \text{LCC}_{\text{WP}})$$

and

$$\text{WCB} = \text{LCC}_{\text{NW}} - \text{WP} - \text{LCC}_{\text{WP}} \tag{9.3}$$

Since the breakeven point for warranty cost effectiveness occurs when WCB = 0, the maximum possible price for the customer to pay may be derived from Eq. (9.3) as

$$\text{WP}_{\text{max}} = \text{LCC}_{\text{NW}} - \text{LCC}_{\text{WP}} \tag{9.4}$$

This framework allows for two cost-effectiveness criteria. Using Eq. (9.1), if the LCC estimate assuming no warranty is greater than or equal to the LCC estimate assuming a warranty, then WCB is positive or zero and the warranty is judged to be cost effective. If $\text{LCC}_{W} > \text{LCC}_{\text{NW}}$, the warranty is judged to be not cost effective, and a waiver can be sought. One might question the cost effectiveness of the warranty when $\text{LCC}_{\text{NW}} = \text{LCC}_{W}$; WCB = 0. It is highly unlikely that the difference would be exactly zero, but could be essentially zero. The reasoning is that if the difference is essentially zero, the weapon system warranty law takes precedence and the warranty is acquired.

Having the warranty price for the with warranty case separated from the LCC to the customer exclusive of warranty price provides additional visibility and a second cost-effectiveness criterion. Looking exclusively at the warranty price, if the warranty price is equal to or less than WP_{max} as calculated from Eq. (9.4) there is a net projected savings to the customer and the warranty would be judged to be cost effective. If the warranty price is more than WP_{max} from Eq. (9.4), the warranty would be judged to be not cost effective, and a waiver should be sought. A more detailed discussion of LCC is found in the section on the CEA process. A key point needs to be emphasized in the LCC analysis framework. The fact that LCC is the only decision metric allows us to consider the LCC analysis as a CEA and therefore provide a cost-effectiveness decision tool. It is desired that the CEA also embrace system effectiveness to provide a more thorough decision tool.

System effectiveness. Ideally, we want to look at the projected impact of the warranty on reliability, maintainability, supportability, and availability to evaluate the system effectiveness portion of the cost-effectiveness figure of merit.

Typically SE is a function of mission reliability, operational availability, and mission capability. For purposes of our discussion, mission capability is not assumed to be affected by the existence or nonexistence of a warranty, and therefore will not be discussed.

Operational availability is basically the likelihood that a system will be ready for use when called on. It is the ratio of up or operational time over a unit timeframe. Mathematically, it is given by

$$A_o = \frac{\text{MTBM}}{\text{MTBM} + \text{MADT} + \text{MLDT} + \text{MTTR}} \tag{9.5}$$

where A_o = operational availability, % – ($0.0 \le A_o \le 1.0$)
MTBM = mean time between maintenance, h
MADT = mean administrative delay time, h
MLDT = mean logistics delay time, h
MTTR = mean time to repair, h

The MTBM represents the uptime, and the sum of MADT, MLDT, and MTTR represents the downtime. The times are clock time. The clock starts when a failure is discovered and ends when the system is restored to operation. Availability is an excellent figure of merit for warranty since it embodies reliability, maintainability, and supportability considerations, all affecting the serviceability of the warranted item.

Mission reliability is the probability that a system will survive a mission of a certain length without failure, given that it was operational at the start of the mission. For a constant failure rate situation (exponential distribution) the mission reliability is given by

$$R_m = e^{-(t/\Theta)} \tag{9.6}$$

where R_m = mission reliability ($0.0 \le R_m \le 1.0$)
e = exponential function
t = specific mission length, h
Θ = system MTBF, h

In cases where the equipment has inherent wearout characteristics, the Weibull distribution is used to estimate R_m. If redundancy is present in the system, the R_m expression becomes more complex, containing the $e^{-(t/\Theta)}$ expression.

Now that we have estimated A_o and R_m, we need to be able to put these two expressions together in some form of mathematical expression.

Since A_o values lie between 0.0 and 1.0, we will define A_o as a probability that the system is ready when called on. Mission reliability R_m is defined as the probability of mission success. Since both A_o and R_m are probabilities, we will form the product of the two and define that as system effectiveness (SE):

$$\text{SE} = A_o R_m \tag{9.7}$$

What we have then is the probability that a system is available and it operates without failure for a certain mission duration. This is a good representation of warranty effectiveness.

We have taken some liberties with the SE derivation. The relationship shown for A_o is most appropriate for a system required to operate continuously, and A_o does not fit the classical definition of a probability. R_m expressed as $e^{-(}t/Theta)$ may be too simplistic a reliability expression. The simplifying assumptions are acceptable when one remembers that we are comparing two warranties. We are interested in the comparative figures of merit for the warranty and no-warranty cases and not the actual magnitude of either figure-of-merit. We are really ranking the two warranties. The comparison made allows us some latitude relative to the analysis methodology.

Cost effectiveness. Now that we have the essential pieces of cost effectiveness, i.e., SE and LCC, we need to put them together.

Figure 9.3 presents a typical CEA outcome.

As the figure shows, we need to be able to evaluate whether SE or LCC is worth more. We will now define SE as the probability of total mission success where the mission includes the availability of the system. From statistical theory related to the binomial distribution, we obtain

$$\mu = np \tag{9.8}$$

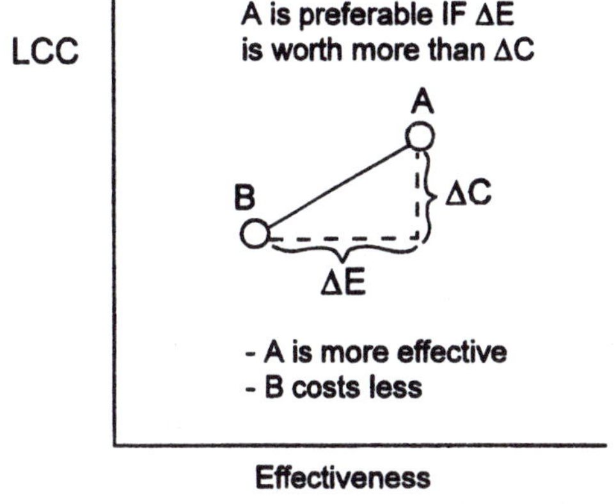

Figure 9.3 Cost-effectiveness analysis outcome.

where μ = mean or expected number of successes
n = number of trials
p = probability of success on a single trial

Using this relationship in our context, we will let p = SE, and n will be the number of missions to be conducted during the deployment period. Then

$$\mu = N(\text{SE}) \tag{9.9}$$

where μ = expected number of successful missions
N = total number of missions conducted
SE = system effectiveness per mission

Incorporating LCC, we have

$$\text{CE} = \frac{N(\text{SE})}{\text{LCC}} \qquad \text{(high is good)} \tag{9.10}$$

or

$$\text{CE} = \frac{\text{LCC}}{N(\text{SE})} \qquad \text{(low is good)} \tag{9.11}$$

CE provides the tool to determine whether ΔSE or ΔLCC is worth more. The units for Eq. (9.10) are expected successful missions per dollar. The units for Eq. (9.11) are dollars per expected successful mission. Either relationship for CE can be used. The only precaution is to remember the convention, i.e., high is good or low is good.

The user should choose the relationship for CE that would be best understood by the reviewers.

Again, the actual magnitude of CE is of no concern. We are concerned with the relativity of the two numbers as substantiation for the buy/waiver decision. The decision maker must decide on the significance of the difference before making the decision.

One last comment before we move on. It could be argued that this approach to CEA requires data that is not known at the time of the decision. It has been my experience that persons knowledgeable of their discipline and the system design can make highly credible estimates of the required input data for both candidates, adequately reflecting the differences. I believe that this technique is feasible and provides the necessary impact of the critical serviceability parameters on the cost-effectiveness decision.

If alternative warranties are offered, the CEA would address each of them as compared to the no-warranty alternative in the same manner as the required warranty.

Other cost elements. In addition to the typical cost elements amenable to mathematical modeling, there are other cost elements which cannot be modeled, but could influence the warranty acquisition/waiver decision. It is recommended that these cost elements be considered only if the CE analysis results are too close to call. These other cost elements include

- *Competition.* A reduction in competition may result if warranty requirements, primarily EPRs, present a high financial risk to the contractors.
- *Breakout.* A decreased opportunity for breakout (provision for certain items as government-furnished equipment (GFE) to the contractors) may occur as a result of warranty application.
- *Default.* Warranty obligations may not be fulfilled because of litigation on liability for system failures or material monetary losses by the contractor.
- *Technology.* Use of advanced technologies in system design may be suppressed if contractors are fearful of potential warranty reprisals and choose older, but proven technology to reduce future system failure risks.

Cost-effectiveness analysis process

Figure 9.4 presents the CEA process.

Form analysis team

It is advantageous at the outset of the CEA process to identify key people in the warranty critical disciplines. These people will be called on to

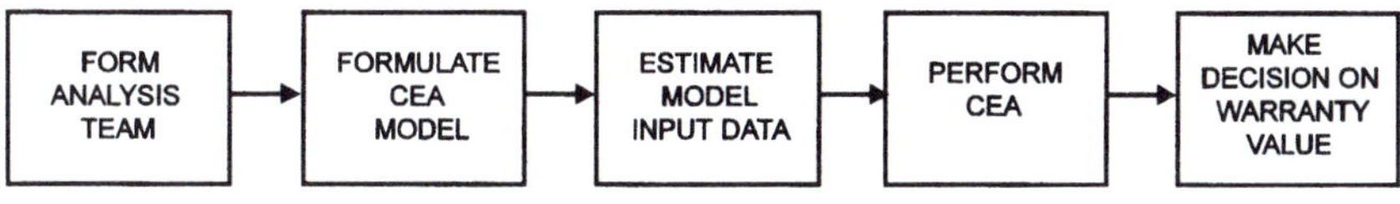

Figure 9.4 Cost-effectiveness analysis process.

represent their disciplines in all CEA effort throughout the program. They do not necessarily have to be located in the program manager's direct sight, but should be available to convene to discuss CEA issues.

It is critical that the team members be knowledgeable in their respective disciplines, especially regarding the impact of warranty requirements on their discipline. For instance, the logistics engineers should be able to impact the flow of warranted items from the field on the required quantity of spares at the depot. It is further critical that the team members be knowledgeable of the system design and operation to facilitate the impact of design changes on the warranty with respect to their disciplines.

The most important function of the CEA team is the estimation of input data to the CEA model. This function will be discussed in more detail in the data estimation step of the CEA process.

Figure 9.5 presents the CEA team.

As Fig. 9.5 shows, all CEA team members are responsible to the program manager. It is important that the program manager have visibility to all the team members to ensure consistency and thoroughness of the CEA. With this organization the program manager has direct access to the CEA analysis data estimators to question them on their rationale. This access allows for a more informed decision by the program manager relative to the acquire/waiver decision. The nature of the data each team member may be required to estimate will be covered in the data estimation step of the CEA process.

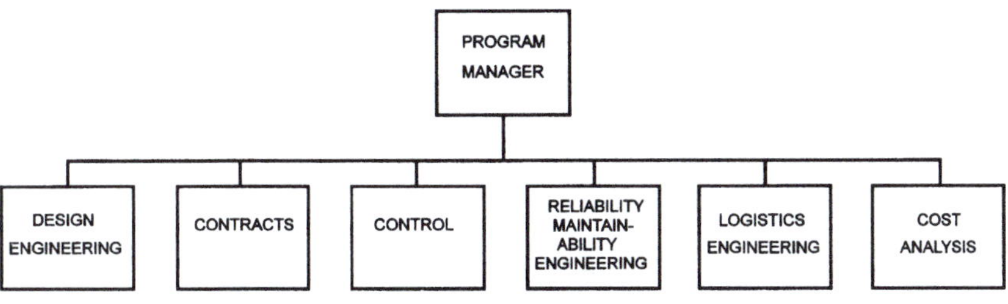

Figure 9.5 Cost-effectiveness analysis team.

Formulate CEA model

We need an evaluation tool to conduct the CEA. It may be possible to use existing models for the CEA, but most likely a tailored model for the particular warranty situation will need to be formulated. It is important that the model complexity be consistent with the degree of definition of available input data. This may require revisions to the model as the program progresses from the demonstration-validation phase to the beginning of the production-deployment phase, when the final CEA is performed.

The model should be as simple as feasible and should be programmed to run on a PC to allow for necessary sensitivity analyses to support the final decision. The model outputs should be easily interpreted to facilitate the justification of the final acquire/waiver decision. Finally, the model should be developed and run exclusively by the cost analyst. From the author's experience, many problems have arisen relative to interpretation of results when too many hands have access to the model.

In discussing the CEA model formulation, we will refer back to the section on CEA concepts, where the theory was developed. The discussion will address LCC first, followed by SE, and finally, CE.

LCC. *Life-cycle cost* is defined as the total cost to the customer (government) of a system over its life cycle. It includes the costs of development, investment, operation, support, and retirement.

We initially express LCC as

$$\text{LCC} = \text{CA} + \text{CS} \tag{9.12}$$

where CA is the acquisition cost and CS is the sustaining cost. Continuing, we have

$$\text{CA} = \text{CD} + \text{CI} \tag{9.13}$$

where CD is the development cost and CI, the investment cost. And

$$\text{CI} = \text{CINR} + \text{CIR} \tag{9.14}$$

where CINR is the nonrecurring investment cost and CIR, recurring investment cost. Also

$$\text{CS} = \text{COS} + \text{CR} \tag{9.15}$$

where COS is the operating and support cost and CR, the retirement cost.

Figure 9.6 presents a typical LCC model structure breaking out the costs to the lowest level. According to Fig. 9.6, each block at the lowest

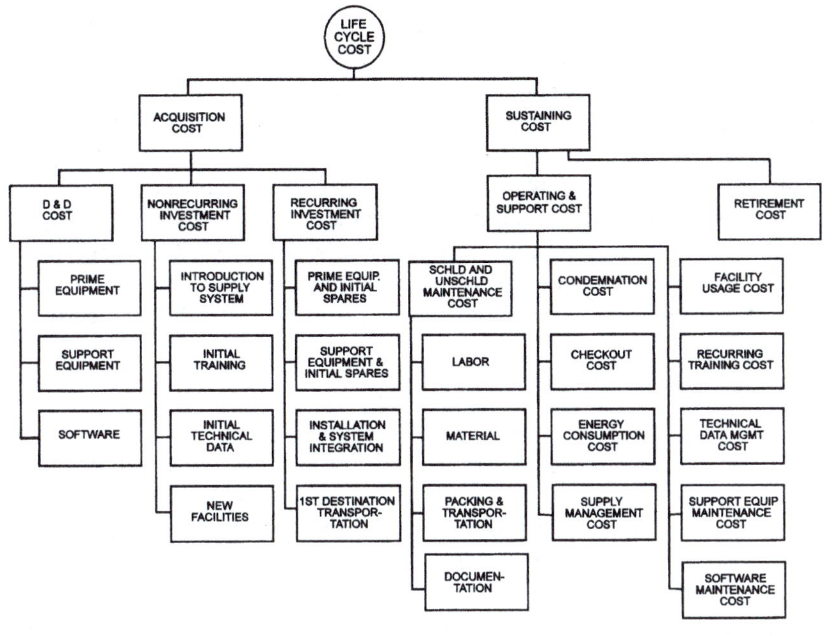

Figure 9.6 Typical life-cycle cost model structure.

level has an algorithm to estimate its cost, and these costs are summed upward to compute LCC per the previous equations.

This same model would be used to calculate LCC for both the nonwarranty and warranty exclusive of warranty price cases. According to Eq. (9.2)

$$\mathrm{LCC_W} = \mathrm{WP} + \mathrm{LCC_{WP}}$$

where WP is the warranty price and $\mathrm{LCC_{WP}}$ = LCC to customer exclusive of warranty price.

Therefore, for the warranty alternative, the LCC model would be used to calculate $\mathrm{LCC_{WP}}$, and the warranty price could be estimated using the techniques in Chap. 7 or as in Eq. (9.4). These two costs to the government would then be added to determine $\mathrm{LCC_W}$. $\mathrm{LCC_{NW}}$ is determined directly from the LCC model. $\mathrm{LCC_W}$ should also include government administration costs and any added cost for transition to an organic depot after warranty. These cost elements are added after the model is run.

System effectiveness. System effectiveness is a function of mission reliability R_m and operational availability A_o. Operational availability is, in turn, a function of maintenance reliability (MTBM), maintainability

(MTTR), and supportability (MADT and MLDT). Care must be taken to be sure that R_m reflects the operational scenario and A_o reflects the maintenance and support concept for the program.

It is not advisable to introduce a special maintenance-support concept for the warranty as this is quite costly. Rather, the warranty should be implemented within the prescribed maintenance and support concept. MLDT is a weighted average based on the probability of having or not having the appropriate spare when needed. The reader is directed to Chap. 5 for a detailed discussion of MLDT.

Equations (9.5) and (9.6) show the A_o and R_m models, respectively, and, per Eq. (9.7)

$$\text{SE} = A_o R_m$$

where SE is the probability that the system is operational when called on, and that the system will last the designated mission length without failure. SE would be calculated for the no-warranty and warranty alternatives using the same model.

Cost effectiveness. Cost effectiveness is a function of LCC and SE. SE is defined to be the probability of total mission success. The expected number of successful missions over the life cycle is a function of the total number of missions to be carried out over the life cycle N and SE as shown in Eq. (9.9).

Incorporating LCC, Eqs. (9.10) and (9.11) show the CE models as

$$\text{CE} = \frac{N(\text{SE})}{\text{LCC}} \qquad \text{(high is good)}$$

or

$$\text{CE} = \frac{\text{LCC}}{N(\text{SE})} \qquad \text{(low is good)}$$

CE provides the model to determine whether ΔSE or ΔLCC, from Fig. 9.3, is worth more. CE would be calculated for the no-warranty and warranty alternatives using the same model. N would be the same for both alternatives.

Estimate model input data

The next step in the CEA process is critical to the success of the process. If the input data to the models does not reflect the differences between the no-warranty and warranty alternatives, the results from the model will likewise not reflect the differences. The CEA team depicted in Fig. 9.5 is called on to provide the input data to the model. There must be adequate time for them to go through the thought process to estimate the data. Some data will not be different between alternatives, or be very close. This is to be expected. The team members must provide sufficient rationale to support the data values to enhance the credibility of the results.

Certain points related to the input data estimation process need to be emphasized. We are asking the team for estimates of the parameters. The sensitivity analysis effort performed as a part of the CEA reduces the importance of the data being accurate since we will be looking at a range of values for the driving parameters. Also, we must continually keep in mind that we are not interested in the accuracy of the CE figures of merit themselves, but rather in the magnitude of the difference between the two. Finally, there is no way to measure the accuracy of the estimates. If knowledgeable people estimate the data it will be reasonable, therefore, the results will be reasonable. Reasonableness is the issue, not right or wrong.

In the author's opinion from experience, this data estimation effort when done thoughtfully by knowledgeable people can produce reasonable data for the CEA model and therefore produce reasonable results. It is achievable.

Referring to Fig. 9.5, the typical data that each team member is asked to provide for both alternatives is shown below.

- *Design engineering.* Development cost.
- *Contracts.* Warranty data on previous similar systems.
- *Control.* Production schedule—unit production cost, and other recurring investment costs.
- *Reliability and maintainability.* Mission reliability, MTBM, MTTR, repair costs, scheduled maintenance intervals, maintenance concept.
- *Logistics.* Initial and replenishment spares quantities, MADT, MLDT, training cost, supply management cost, data management cost, support concept.
- *Cost analysis.* Standard service cost factors.
- *Program management.* Operational scenario, details of the warranty requirement.

This list of typical input data provided by the team members is by no means complete, but serves to indicate the type of data required from each team member and the degree of detail required for the analysis.

There can be no shortcuts through the input data estimation effort. The effort demands diligent monitoring by the program manager and thoughtful attention to detail by the team members.

Perform CEA

We now have a viable computerized CEA model and input data estimates reflecting the differences between the no-warranty and warranty alternatives. What remains is to perform the CEA.

Figure 9.7 presents the LCC analysis process.

According to Fig. 9.7, there are three major types of inputs required by the model: doctrines, system characteristics, and standard factors. The doctrines include the procurement plan, the operational scenario, and the maintenance-support plan. Establishing a baseline for these three concepts is essential before any LCC analysis can be performed. The operational scenario provides the distribution of the systems, the usage rates of the equipment, and the number and location of operational bases. The maintenance-support concept defines the number of maintenance levels, what type of maintenance is performed at each level, and the sparing concept. The procurement plan gives us the system quantities, and the production and deployment schedules.

The system characteristics include such data as the MTBF, MTBM, MTTR, unit production cost, and weight. This data comes from the team and represents the design of the system. All this data is contractor-controllable.

The standard cost factor data comes from the services and includes such things as maintenance personnel labor rates, shipping costs, line item entry and retention costs, and training costs. These cost factors are updated by the services and should reflect the current year.

This data combines into an input database. The data is input to the model. Notice that the doctrines influence the formulation of the model. For example, if the procurement schedule is staggered rather

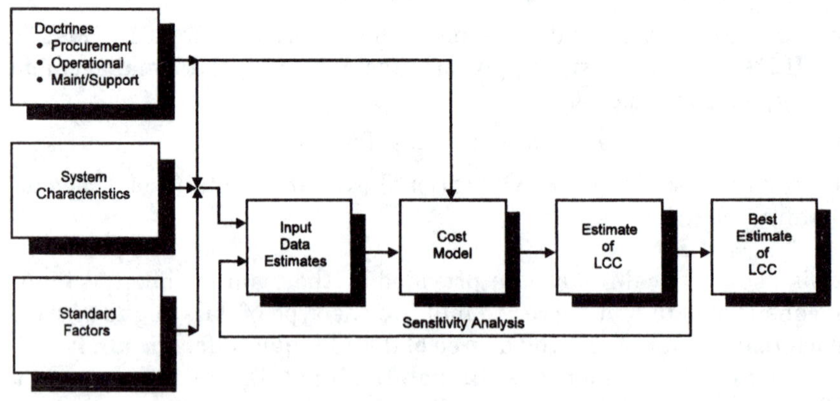

Figure 9.7 Life-cycle cost analysis process.

than all in one year or if the maintenance concept directs two levels of maintenance rather than three levels, these considerations affect the format and content of the model.

The computerized LCC model is exercised for both the no-warranty and warranty alternatives to provide a baseline estimate for both. Key input parameter drivers such as MTBF and unit production cost are varied within reasonable ranges and the effects on the bottom-line LCC noted. This sensitivity analysis allows us to determine a reasonable range of values for LCC for both alternatives and identifies the threshold values of the drivers indicating the breakeven point between the two alternatives. The best estimate of LCC occurs when the maximum benefits of the sensitivity analysis are realized and a reasonable range of LCC can be estimated.

Coincident with the comparative LCC analysis, the comparative values of A_o and R_m are determined from the team data and the comparative values of SE are calculated per the given model, and sensitivity analyses run.

What remains is to calculate the comparative values of CE. The operational scenario provides us with the mission frequency from which we can calculate the total number of missions over the life cycle. Using the sensitivity analysis information from the LCC and SE analyses, we can perform sensitivity analysis on CE to determine the robustness of the results to changes in the input data drivers. Having used the most likely values of LCC and SE for CE and then having performed the necessary sensitivity analyses on CE, we are now ready to make a decision on the value of the warranty.

Make decision on value of warranty

The program manager's decision is based on the results of the baseline CEA and the associated sensitivity analyses. It is not likely that the decision is robust to any values of the input data drivers. The threshold values of the drivers indicating at what CE values the breakeven point occurs are therefore useful information. If the drivers can truly be that bad or that good, then the team's input as to most likely value will help bound the problem and allow for a decision. For instance, a lower threshold on MTBF might be 200 h. If the MTBF is 200 h or less, the warranty alternative is preferable; if MTBF is greater than 200 h, the no-warranty alternative is preferable. The program manager needs to question the reliability engineer as to the likelihood of the MTBF being either less or more than 200 h. Then the program manager's decision is based on the reliability engineer's opinion. There may be other driver thresholds that need to be considered in the decision. These are evaluated in the same manner as was MTBF.

It should be pointed out that management judgment is a major part of the decision process. The CEA is a tool to aid that decision and not replace management judgment. The manager will have to decide whether a difference between the CE values is significant. There is no rule of thumb. The manager must make that call on the basis of the information available and personal judgment. If there is no significant difference between the two alternatives, the manager is led to consider other factors such as competition, default, or technology as discussed in a previous section.

If the decision is that the warranty is cost effective, administrative details are finalized and the contractor implements the warranty on production units. If the warranty is not cost-effective, the program manager independently compiles results and supporting rationale and begins to implement the waiver process.

A simplified CEA with the associated decision rationale is presented as an example at the conclusion of the chapter.

Warranty Waivers

Per Sec. 2403 of the Title 10 U.S. Code, warranty waivers can be sought in cases where the warranty, if implemented, would not be cost effective for the government.

Figure 9.8 shows the basic warranty cost-effectiveness process.

As depicted in Fig. 9.8, the warranty CEA is performed by the customer as described in the previous section and a decision made relative to the cost effectiveness of the warranty. If the warranty is cost effective, the administrative details are finalized and the negotiated warranty is implemented by the contractor. If the warranty is not cost effective, the armed services should petition the Secretary of Defense

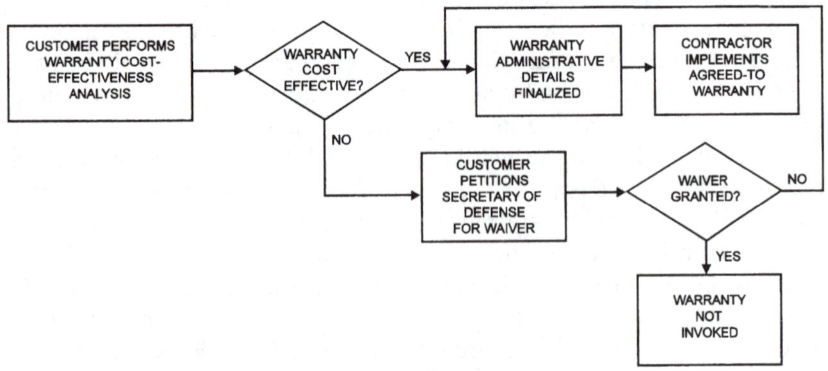

Figure 9.8 Cost-effectiveness process.

for a waiver. If the waiver is not granted, then the negotiated warranty is implemented, anyway. If the waiver is granted, then the warranty is not invoked on that program.

As was pointed out earlier, historically, very few waivers have been requested to date. The major reason for this was determined by the GAO to be that the services were not performing CEA or were not performing them adequately. For warranties to be valuable to the services, they must be cost effective. If they are not cost effective and waivers are not sought, then their implementation represents a bad business decision for the government—more specifically, they are wasting their money. The writers of the warranty law never intended that every program's warranty would be cost effective and therefore implemented; that is why they included the waiver option for warranties that are not cost effective. The answer to this dilemma lies in the diligent implementation of CEA to provide credible rationale to support waiver decisions.

The Army has taken a firm stand; to wit, they are going to purchase only those warranties that adhere to the law and make good business sense.

Example

Following is a simplified example of a CEA with the associated decision rationale. The basic information is as follows:

- *Warranted item.* F-33 aircraft radar subsystem.
- *Warranty type.* Failure-free, with MTBF guarantee.
- *Warranty requirements.* Materials and workmanship, design and manufacturing essential performance on MTBF.
- *Warranty coverage.* Individual item—24 months; systemic—60 months.
- *Warranty remedies.* Repair or replace failures; redesign or retrofit defects; free spares for MTBF guarantee deficiencies.

Figure 9.9 presents the CE factor interactions for the CEA.

The LCC analysis is run with the following model:

$$\text{LCC} = \text{CD} + \text{CI} + \text{COS} \qquad \text{(same notation as before)}$$

The results are tabulated below using previous notations.

$$\text{LCC}_{\text{NW}} = \$2.5 \text{ million} + \$122\text{million} + \$147.5 \text{ million} = \$272\text{million}$$

$$\text{LCC}_{\text{WP}} = \$3.1 \text{ million} + \$127\text{million} + \$146.0 \text{ million} = \$276.1 \text{ million}$$

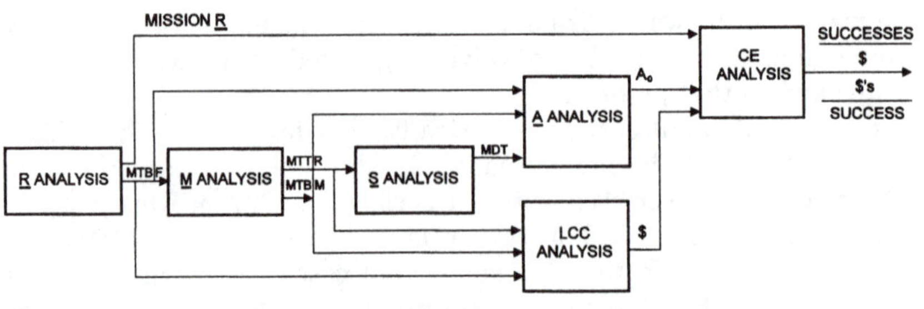

Figure 9.9 Cost-effectiveness factor interactions.

and

$$\text{WP} = \$13.9 \text{ million}$$

where WP includes government administration cost and transition cost. Therefore

$$\text{LCC}_{\text{W}} = \$13.9 \text{ million} + \$276.1 \text{ million} = \$290 \text{ million}$$

The ΔLCC is then = $18 million.

Sensitivity analysis indicates MTBF to be the major LCC driver. The SE analysis is run with the following model:

$$\text{SE} = A_o R_m$$

The results follow:

$$\text{SE}_{\text{NW}} = .9505 \times .9802 = .9317$$

$$\text{SE}_{\text{W}} = .9470 \times .9850 = .9328$$

The ΔSE is then 0.11 percent. Sensitivity analysis indicates that MTBM (derivative of MTBF) is the major SE driver.

Now that we have the LCC and SE values, we can move on to the determination of the comparative values of CE.

The CE analysis is run using the following model:

$$\text{CE} = \frac{\text{LCC}}{N(\text{SE})} \qquad \text{(low is good)}$$

where N is determined to be = 1,440,000 missions. The results are tabulated below:

$$\text{CE}_{\text{NW}} = \$203 \text{ per successful mission}$$

$$\text{CE}_{\text{W}} = \$216 \text{ per successful mission}$$

Sensitivity analysis is run on CE using the information from the LCC and SE sensitivity analyses. MTBF is allowed to vary ±25 percent, and MTBM is allowed to vary ±25 percent. In this case, 75 h is the lower MTBF threshold value where the CE values are equal, and 45 h is the lower MTBM threshold value where the CE values are equal. On questioning the reliability engineer, the program manager becomes quite certain that the true MTBF value will be >75 h and the true MTBM value will be >45 h.

The program manager is satisfied that the true MTBF and MTBM values should comfortably exceed the threshold values and further that the CEA results are robust within the constraints given above.

The program manager is prepared to make the decision, but decides that it is necessary to be able to explain the rationale for the decision to the boss (program manager's immediate supervisor), who is very practical. The program manager chooses the no-warranty alternative because of the $13 lower cost per successful mission, and notes that the total savings over the 1,340,000 successful missions is approximately $17.5 million and these savings are approximately 6 percent of the total radar subsystem expected LCC. The program manager feels this is adequate backup to support the significance of the difference between the two alternatives and, having informed the boss of the decision that the warranty would not be cost effective, organizes the material and proceeds to request a waiver.

Finally, the 0.11 percent higher SE with the warranty alternative was overshadowed by the $18 million difference in LCC favoring the no-warranty alternative. In this example, ΔLCC was worth more than ΔSE.

A final note before we leave Chap. 9. Situations may arise where the method presented herein is not feasible. If that is the case, then use a technique that is feasible. A pure LCC analysis is far better than no analysis, even though the effectiveness factor is not addressed. If the effectiveness factors cannot be treated quantitatively, then cover them qualitatively. The point is that the effectiveness portion should not be ignored and should have some bearing on the final decision on cost effectiveness. The bottom line is to perform the CEA to the degree feasible, with valid rationale to support that analysis.

Chapter

10

Warranty Implementation and Administration

Staffing

Warranty administration is a full-time job. It takes not only a full-time person to administer a warranty program but also a competent staff of support personnel, working either directly or indirectly for the warranty manager, to properly implement the warranty. The size and quantity of staff depends, of course, on the size of the program, the amount of equipment to be monitored, and the data-reporting requirements, as detailed later.

The primary tasks of the warranty administration manager are to

- Review the warranty contract and make recommendations to managers as to the implementation process needed.
- Set up the actual implementation program, procedures, and data responsibilities.
- Keep an ongoing company liability status.
- Be the company interface between your company and the customers.
- Make repair-cost-fault determinations.

The warranty manager should work closely with other functions in the company such as contracts, procurement, accounting, sales, repair, and shipping, because all these functions are affected by the warranty.

The staff size is dictated by the following elements:

- Whether you are the prime contractor or a subcontractor. The prime contractor, whether a civilian organization or a Department of Defense contractor, has the major responsibility for the warranty, and must assume the role of the person coordinating the program. Subcontractors usually have much less warranty requirements, and less responsibility than the prime, and their staffing size is less in proportion. However, at least one individual at each subcontractor facility should be responsible for insuring that the warranty requirements are monitored, tracked, and reported.
- The amount and value of the equipment usually dictates how many people are needed for administering a warranty program. A small program could possibly be administered adequately by one person with a desktop computer. In many commercial companies this task is assigned to someone who usually provides another function, such as the service manager at an auto dealership. Whoever performs this function should be someone who is involved with the repair process, and has knowledge of the product from the time it arrives until it is shipped out or completed repair.
- The amount of data reporting requirements also adds to the amount of people needed for warranty administration. If there are weekly or monthly reports required, then it takes more people to administer it than it would if there were no data reporting requirements. Typists, clerks, and others add to the size of the staff for most companies with warranty requirements.
- How warranties and warranty requirements are perceived in a company probably dictates the size and importance of a warranty team more than anything else. If upper management perceives the warranty as a valuable asset for sales and marketing, then a larger warranty staff will be in place and more emphasis put on it. If the company regards the warranty as one more thing to cause increasing costs, and considers it just another necessary evil to have to perform because it is required, then it will probably be done by someone who has other tasks as their primary job, and will handle warranty situations as the need arises. Too often this is the case, and a major opportunity for profit is lost.

Asset Tracking

Flow

Assets under warranty must be tracked, whether they are commercial products, or built on a government contract. The requirements, however, vary greatly, depending on the customer. Most commercial products are serialized, and tracked from production bases, such as television sets and computers. The prime responsibility lies with the seller of the product, and the warranty begins from the day of sale. In some cases a defective product will be returned to the manufacturer, but in other cases, will be returned to the seller. Each process has its own merits, has been in place for decades, and is fairly well established in the industry. However, the relatively new warranty law in regard to government contracted equipment has imposed additional requirements on defense contractors, and has injected new tasks into their programs. Figure 10.1 depicts the flow of a government contractor's tasks for a typical equipment procurement:

Requirement received. The initial request for proposal (RFP) or request for quote (RFQ) is assumed, for this illustration, to have been completed successfully, and a contract is now received with a warranty requirement, which must be evaluated as a part of the contract.

Assignment of person or team. Many companies have a warranty department established which handles all contract warranties. If that is the case, then it is assumed that procedures are in place to handle the

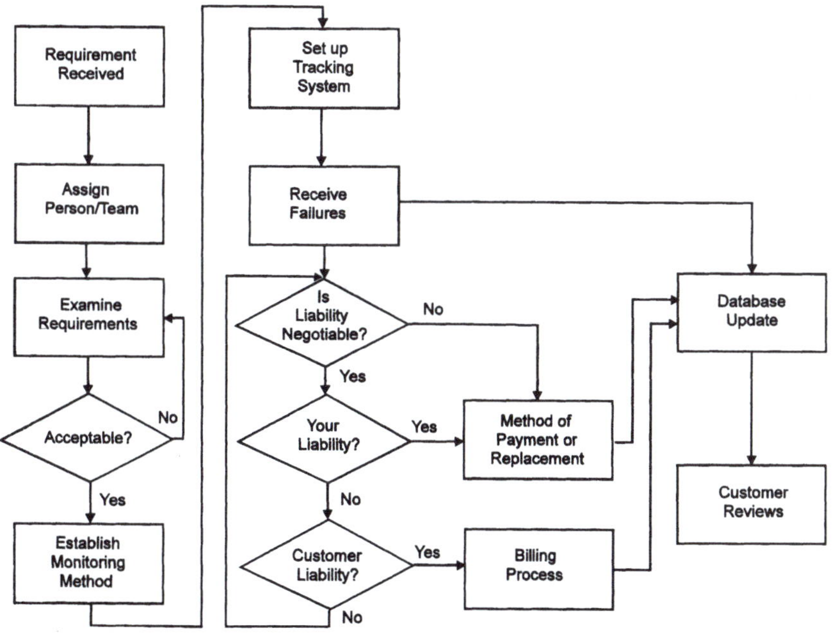

Figure 10.1 Implementation process flow diagram.

remainder of the implementation process. The flowchart shown in Fig. 10.1 is provided for those companies where the warranty is handled by the program, and may or may not have experience dealing with the many tasks that must be set up. A warranty manager must be assigned, who must then relegate responsibility for inputs to implement the warranty to selected personnel.

Examine requirements. The assigned warranty team first examines the requirements to see that they are compatible with the company's capabilities and objectives. Are the requirements achievable? If not, what exceptions will be communicated to the customer? Are exceptions allowed? These questions must be answered before further action can be taken.

Is warranty acceptable? The warranty manager must examine the warranty requirements of the contract to determine whether the requirements are acceptable and achievable. They can determine this only by discussing the technical aspects of the program with the technical staff.

Establish monitoring method. Once the warranty is acceptable, a method of monitoring the warranted items must be established. This can be accomplished at the production facility by setting up a warranty database to monitor the production flow, quantity, and serial numbers. The database personnel usually work closely with the configuration department, since they have a common goal.

Set up a tracking system. Tracking the warranted assets is the most difficult and costly part of the warranty program. It requires close communication with the customer at field locations, and quite often, company technical representatives at the customer's locations. The capability must be available to track each asset to its current location, because future retrofits, recalls, and so on require that the locations of the equipment to be recalled be known.

Receive failures. When production is in process, and sufficient time has passed for units to start failing in the field, the units will begin to come back for repair. Adequate procedures must be in place to handle all aspects of the repair of the returned systems.

Is liability negotiable? Is the liability for the cost of repair of returned units already established by the warranty, or is it negotiable, due to the nature of the failure? If it is not negotiable, the procedure for handling the method of payment or replacement must be set up. If it is negotiable, then, for customer liability, a method of billing must be set up with the accounting department. In either case, the warranty database must be updated to reflect the action, and to provide data for customer review. This record also provides the company management with the cost of warranty at any given point in time.

Customer reviews. It cannot be emphasized strongly enough that frequent reviews with the customer as to the status of the warranty are the best method available for establishing customer rapport. A good working relationship between the seller and the buyer can produce a more workable warranty for both current and future contracts.

Supplier and customer obligations

It is the customer's obligation to provide in the requirements a workable warranty which will provide them with the visibility and coverage they require, without imposing unnecessary hardships on the supplier. That is often more difficult than it sounds. It is only human nature to get the most you can out of any endeavor, and quite often this author has seen warranty requirements which were unreasonable. Many negotiations were necessary to establish a workable warranty which would suit the customer. However, the customer needs to understand that the supplier cannot provide a warranty which may make it extremely unprofitable to execute. Warranties, in some cases, are considered free, but they still produce costs for the supplier.

It is the supplier's obligation to provide a warranty on the product which will satisfy the customer, and provide a degree of confidence that there will be no additional costs for procuring this asset. Companies who build government equipment obviously are proud of their products, and are willing to back them with a reasonable warranty, and few, if any, regard a warranty clause as an obligation they are not willing to accept. Most government contractors have provided a warranty on their products for the past 20 or 30 years, even before the warranty law went into effect, just to ensure that their products operated in the field to the satisfaction of the customer.

Identification of what assets to track

Identifying exactly which assets will be tracked seems almost like a moot question. All the assets. Right? Wrong. The complexity of modern-day equipment has produced many assemblies and subassemblies within each box, and some major assemblies may contain hundreds of printed-circuit boards, chassis-mounted components, and so forth. To try and track every one of them would produce an unworkable data tracking system of enormous size. The cost would be prohibitive. Most tracking systems the author is familiar with go down to serialized subassemblies only. Many large systems are tracked only to the top assembly, and all subassemblies within that assembly have the same warranty as the top assembly. This presumes, however, that the top assembly is returned whenever a repair is necessary. If not, then tracking will have to be performed to the lowest level of subassembly which may be sent back for repair. The supplier and the buyer need to work out a plan which is acceptable to both.

Database Management

Computer hardware and software

The most powerful tools available in monitoring warranty compliance are the computer database and its supporting data collection system. It can be broken into four major areas:

- *Type.* It must be programmable, but can be as small as a PC.
- *Size.* It must provide adequate memory for total production, plus future updates and additions.
- *Capability.* Networked on-line databases are ideal and give the greatest amount of information with least redundancy.
- *Operation.* Select hardware and software with ease of inputs and outputs, and for updates.

Inputs and outputs

The inputs from Figure 10.2 consist of the following:

- Original configuration data
- Reconfiguration data
- Field operational data
- Maintenance and repair data

The outputs (Fig. 10.2) consist of

- Repair reports (logistical data)
- Customer reports (data items)
- Failure reports (reliability data)
- Management reports (internal data)

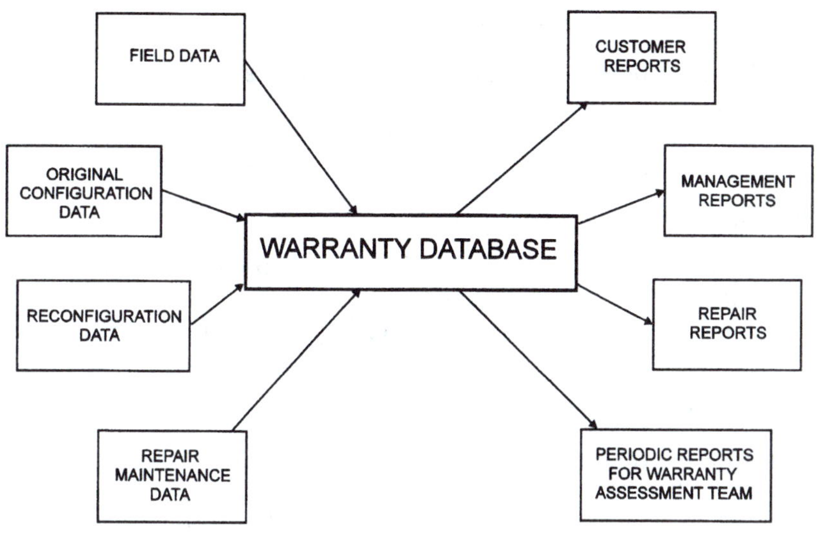

Figure 10.2 Warranty database.

Procedures

The procedure for database implementation can be summarized in seven steps:

- Determine key data elements.
- Analyze existing databases and systems for adequacy.
- Evaluate future requirements for other uses or projects.
- Make cost tradeoffs of performance requirements versus cost.
- Select hardware and software which meets requirements.
- Develop the I/O tools: sources of inputs and types of outputs.
- Continue maintenance, updating, and programming.

A more detailed analysis of the summary given above is provided for each of the seven steps:

- Determine key data elements:
  - The key data elements determine the how, why, when, and where of data collection methods.
  - Hidden monitoring costs increase in proportion to the number of data elements.
  - Examples are operating hours, frequency of returns for repair, serial numbers, sold date, and customer.

- Analyze existing databases:
  - Create a new method only if the existing databases are not adequate; evaluate available resources.
  - Keep the number and complexity of data collection items to a minimum.
- Evaluate future requirements:
  - Do not create a manual system dependent on manual keypunch only. As quantity increases, cost will, too.
  - Wherever possible, automate or tap into other databases. Be creative!
- Perform tradeoffs of performance versus costs:
  - Forecast computer processing and storage requirements.
  - Consider the speed, size, and storage capacity necessary.
  - Compare costs of available hardware and new equipment.
- Select hardware and software which meet the requirements:
  - Select hardware which can be used throughout the life of the product's warranty. Selecting a cheap system which becomes inadequate during the life cycle of the warranty program can be more expensive in the long run.
  - The recommended software is an existing or preprogrammed database management system to reduce the number of people needed to set it up.
  - Select software which allows for easy report generation.
- Develop the input/output tools:
  - *Inputs into the database.* Data collection and data entry can be accomplished simultaneously by manual keypunch entry while automated inputs are automatically being fed.
  - *Outputs from the database.* The types and quantity of output data and reports available are limited only by the imagination and creativity of its creator.
- Maintain the system. Programming changes become a way of life throughout the warranty period. New report requests, retrofit activity and contract modifications make additional programming necessary for the life cycle of the warranty period.

An amazing fact seems to be that if you produce a few good reports, you will be asked to generate many more reports, saying the same thing many different ways.

Examples of databases

The type and configuration of databases and their outputs are limited only by the imagination of their creators. Any amount of data may be presented, in any number of formats. Tables 10.1 and 10.2 are examples of data produced by a database which tracks a production program of up to 500 units per month, and the return and repair of approximately 10 units per month. The illustrations shown are for example only, and do not reflect any particular company's operation, but exhibit data which would be useful to management.

Repair Considerations

Demand on supplier resources

When manufacturers produce a product, one major consideration is the expense (or profit) of those units when they fail and must be repaired. There are many options as to how that can be accomplished. The repairs can be made at the factory, using already existing facilities that produce the product, use the seller's facilities at other locations, or set up a repair facility specifically for repair or rework of returned units. Many questions have to be answered before the final decision can be made. Who determines liability when units are received? Are repair procedures adequate at other locations? Will training be required at multiple locations? Does the same warranty apply when repairs are made at another location? Can you assure the quality performed at other repair locations? Can a unit repaired at another location come to your location for the next repair? Various solutions have been discussed between the author and government agencies or contractors, and apply to a variety of types of companies and their products:

- Your facility performs all warranty repairs, and other locations perform only nonwarranty repairs.
- The warranty on units applies only to the first repair, then is voided.
- No workmanship warranty on repairs performed at other locations, but the rest of the warranty is still valid.

If another location is to perform repairs on your warranted product, the following is a recommended guide:

- A systematic, organized procedure must be set up to handle all the different types of failures which may arrive.
- For government contractors, a bonded stockroom must be established for receiving of returned units.

TABLE 10.1 Example of Warranty Database

Extension no.	Date extended	Serial no.	Type	Customer	Repair TAT
1	October 16, 1990	1034	C	USN	23
2	November 7, 1990	1545	C	USN	14
3	November 15, 1990	1737	C	USAF	8
4	November 24, 1990	1656	C	USN	80
5	December 5, 1990	1448	S	USAF	12
6	December 13, 1990	1867	S	USAF	15
7	January 6, 1991	1745	C	USAF	15
8	January 15, 1991	1954	S	USN	16
9	January 17, 1991	1546	C	USAF	32
10	January 19, 1991	1475	S	USAF	16
11	February 5, 1991	1768	C	USN	9
12	February 9, 1991	1687	C	USAF	16
13	February 18, 1991	1473	S	USAF	25
14	February 23, 1991	1794	S	USN	38
15	March 3, 1991	1672	S	USAF	78
16	March 14, 1991	1864	C	USN	121
17	March 25, 1991	1673	S	USN	13
18	April 13, 1991	1755	C	USAF	33
19	April 27, 1991	1464	C	USN	22
20	May 16, 1991	1546	S	USAF	13
21	May 18, 1991	1355	C	USN	15
22	May 27, 1991	1568	S	USN	16
23	May 29, 1991	1477	C	USAF	18
24	June 8, 1991	1665	S	USAF	19
25	June 16, 1991	1779	C	USN	5
26	June 24, 1991	1968	C	USN	21
27	July 3, 1991	1886	C	USN	15
28	July 12, 1991	1579	S	USN	16
29	July 23, 1991	1348	S	USAF	17
30	August 2, 1991	1432	C	USN	33
31	August 13, 1991	1243	S	USAF	22
32	August 22, 1991	1434	C	USAF	15
33	August 25, 1991	1342	C	USN	11
34	September 7, 1991	1564	C	USAF	66
35	September 16, 1991	1455	C	USN	35
36	September 28, 1991	1676	S	USN	13
37	October 7, 1991	1565	C	USAF	16
38	October 12, 1991	1687	S	USN	18
39	October 16, 1991	1676	C	USAF	34
40	October 24, 1991	1488	S	USN	13
41	October 27, 1991	1577	C	USAF	16
42	November 4, 1991	1699	C	USN	15
43	November 6, 1991	1588	C	USN	17
44	November 15, 1991	1753	C	USN	19
45	November 17, 1991	1664	C	USN	22
46	December 3, 1991	1855	S	USAF	15
47	December 8, 1991	1746	S	USN	12
48	December 15, 1991	1056	S	USN	18

TABLE 10.2 Example of Warranty Database

Serial no.	Customer	Contract no.	TAT	Verified failure?	ETM Sale	ETM In	ETM Out
1332	USN	FY88	21	Y	15	23	36
1443	USN	FY89	17	Y	13	19	23
1352	USN	FY88	22	Y	7	51	58
1643	USN	FY89	23	Y	13	56	64
1544	USAF	FY89	7	Y	22	33	40
1755	USAF	FY89	9	N	10	10	17
1673	USN	FY89	15	Y	8	60	66
1566	USN	FY89	24	Y	12	20	26
1685	USN	FY89	101	Y	15	101	155
1564	USAF	FY89	23	Y	15	55	66
1367	USN	FY88	21	Y	10	60	70
1255	USN	FY88	31	Y	13	65	88
1346	USN	FY88	21	Y	19	122	140
1435	USN	FY88	15	Y	16	105	123
1344	USAF	FY88	16	Y	19	88	94
1656	USAF	FY89	16	Y	11	19	35
1545	USN	FY89	17	Y	17	35	52
1778	USN	FY89	24	Y	18	20	44
1566	USN	FY89	15	Y	17	35	60
1327	USN	FY88	12	Y	20	151	166
1432	USAF	FY88	11	Y	13	70	82
1324	USAF	FY88	17	Y	12	62	77
1243	USN	FY88	62	N	16	201	210
1435	USN	FY88	35	Y	15	87	97
1364	USN	FY88	25	Y	21	75	89
1555	USN	FY89	16	Y	13	62	71
1414	USN	FY89	18	Y	12	44	55
1326	USAF	FY88	28	Y	19	90	98
1635	USAF	FY89	35	Y	23	65	78
1528	USN	FY89	28	Y	13	63	79
1737	USN	FY89	44	Y	15	80	101
1646	USN	FY89	36	Y	12	62	88
1532	USN	FY89	26	Y	13	25	35
1624	USN	FY89	17	Y	15	55	65
1723	USN	FY89	18	N	16	16	43
1836	USAF	FY89	36	Y	11	22	45
1245	USAF	FY88	55	Y	18	123	145
1334	USAF	FY88	17	Y	12	102	120
1466	USN	FY88	28	Y	13	75	85
1555	USN	FY89	12	Y	21	33	44
1471	USN	FY88	22	Y	16	62	77
1666	USN	FY89	54	Y	21	32	62
1544	USN	FY89	33	Y	17	65	75
1717	USAF	FY89	25	Y	18	70	81
1751	USAF	FY89	17	Y	20	30	41

- Incoming inspection must be performed on all returned units, preferably with the government representative present.
- Pertinent parameters should be noted [expiration dates, elapsed-time meter (ETM) readings, date of last return, etc.].
- Determine liability before the repair begins (if possible).
- Replacement of configured items must be updated in the database.
- Determination of whether and how a new warranty expiration date is required, and who is to mark the unit.
- Determination of who pays for questionable liability repairs, how, and when.

Impact on production

If your production facility is going to be used as the primary repair site (depot), all areas of the company are affected by the warranty and the associated injection of returned units for repair. The following is a detailed description of a typical government contractor's production or repair flow, as shown in Fig. 10.3, where the production facilities are also used for repair of returned units.

Vendor subassemblies and parts. The majority of prime contractors do not build all the assemblies and subassemblies themselves; rather,

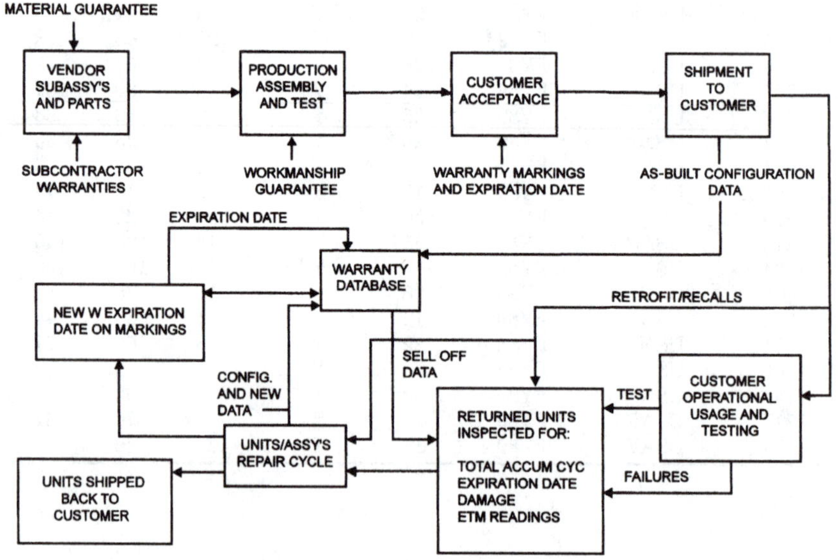

Figure 10.3 Equipment warranty flow.

they buy many of the parts and material from subcontractors. These parts should have a subcontractor warranty on them for at least the duration of the warranty on the prime equipment. The material purchased has a material warranty on it which has to be monitored and recorded. Someone at each end of the process (seller and buyer) must be responsible for warranty actions. If at all possible, the warranty manager should participate in the procurement process. Warranty clauses quite often are included in the purchase order. Subcontractors should be required to keep test results, in case of future problems. Subcontractor warranties also reduce the liability of the prime contractor. Retrofit costs incurred by subcontractors sometimes can be passed on to that supplier if the proper documentation has been recorded during the procurement.

Production assembly and test. During the actual building and testing of the product, the workmanship portion of a warranty comes into play. It is the most difficult part of warranty obligation to control. Pride of workmanship must be instilled into production workers. The Japanese have led in this field for many years. It is beginning to show up in American production, because the reality of more profit from less repairs and returns has shown that a better-built product will produce greater profit. The primary concern of meeting production schedules is giving way to quality considerations and providing a better product, which produces greater future economic returns.

Customer acceptance. On customer acceptance, being bought by the customer, the equipment must be marked with warranty duration, expiration, and manufacturer data in accordance with MIL STD 129 and 130. Lag time between sell-off and delivery sometimes becomes a problem, and procedures must be established to determine the appropriate dates and markings.

Shipment to customer. When the product is ready for shipment, the configuration of the equipment, as-built configuration data, must be provided to the configuration and warranty databases. It is the only time the actual configuration of the final product is established, and is invaluable as to its need when equipment is returned for repair, or has to be identified down to the subassembly level for possible defects for retrofits.

Customer operational usage and testing. Sold units delivered to the customer undergo normal usage, or in some cases, special tests. Whether they are returned because of a failure, or because after a certain amount of time they must be retested, they will come back as units under warranty. They should be segregated from units which may be returned that are not under warranty. At government contractor facilities, they usually come into what is called a "bonded stockroom," which denotes that they are government property and must be handled as such.

Inspection

The returned units must be physically inspected for any damage because the exceptions clause of most warranties does not cover customer-induced damage. Depending on the type of equipment, they should also have the warranty expiration date logged in, as well as any readings on elapsed-time meters, or in the case of aircraft engines, the total accumulated cycles. Sell-off data from the database should be pulled up to compare data at sell-off time with the current data.

Marking

The beginning and end dates of the repair cycle should be recorded on all units, so that the turnaround time (TAT) for repair can be calculated. Some companies do not require this information, but it is good managerial data. All repairs made to a unit should be logged, and if necessary, any changes or replacement of serialized assemblies should be updated in the configuration database. Many contracts require a new warranty expiration date on units returned to the seller, and the shipping department must be notified of any marking changes needed. The warranty database must also be updated with the new expiration date.

Shipping

When repairs are completed on a returned unit, it should be treated as a new delivery, because it may come back again as a failed unit. The shipping department must have an established procedure for packing and crating the units, and must ensure that all containers are marked properly before being shipped. The shipping documents must be relayed back to the warranty department, so that the warranty database can be updated with the shipping and delivery dates.

Reporting

Status reports

Most commercial companies that this author has dealt with do not prepare formal warranty reports. The few who do, for example, Sears repair centers, report only the repair or replacement cost. Tracking of assets is generally not necessary, since the warranty expiration date is stamped on most commercial products, or are from the date of sale. However, Mercedes-Benz of North America keeps all repairs performed on their cars at the dealerships in one central computer system, and can call up the repair history of any of their automobiles on request. Other car manufacturers may also be doing this now, but the few American auto dealers this author contacted did not.

The status reports required on government contracts are many and varied in content. The procuring agency generally has its own list of reports shown on the DD1423, and a list of these would be extensive, and probably incomplete, because of the different requirements of the various agencies. A copy of the agencies' lists can be procured from the government at no cost simply by asking for them in writing, and having a need to know. The following are reports which are generated by most major contractors to stay current with the warranty program:

- Total production quantity status
- Quarterly production quantity status
- Units on hand for repair status
- Total units repaired status
- Failure analysis reports
- Projected failure rate on production contract
- Contract revisions to warranty status
- Average cost per unit for repair
- Total cost to date of repair or replacement
- Subcontractor(s) repair or replacement status
- Cost of warranty status (both annually and cumulative to date)
- Projected worst-case liability status

Cost reports

Cost reports for commercial products have been discussed in the preceding paragraph, and will not be reiterated here. The cost reports on government contracts present an entirely different set of problems. The number of required cost reports depends mostly on that type of contract on which the production units are being built. On firm-fixed price (FFP) contracts, the cost of the program has been agreed upon, and whether the actual cost exceeds or underruns the negotiated price is only known in detail by the seller. The government has the right, as they should, to ask for current warranty costs from the contractor, but actually the contractor is not obligated to divulge this information. If the program is running as was anticipated, most contractors will provide a degree of insight into their costs to maintain good customer relations. It must be remembered, however, that at the completion of even a FFP contract, the government has the right, and usually does, to audit the contract, and the actual costs will be known. This gives the government insight into future production costs, and will be used for future negotiations.

Cost-plus contracts have several variations, such as cost-plus fixed fee (CPFF) and cost-plus incentive fee (CPIF). But regardless of the type of cost-plus contract, cost reporting is rigid and regulated. The most common reporting method used currently is the cost and schedule control system (C/SCS), which monitors budgets and costs against the quantity and schedule of the units produced. It is very exacting, and must be reported at least monthly. C/SCS gives good insight into the current cost status of a program, as well as into the eventual total cost. Many cost status reports are required to satisfy this type of contract, and vary from agency to agency. They are specified in the contract data requirements list (CDRL) of any particular production program, and a price has been negotiated for the data prior to contract award. Any deviations to the original cost estimate must be renegotiated as actual costs accrue, especially if there is any cost increase.

Review board

Review boards for warranty generally pertain only to government programs. This author has not been able to find any evidence of a commercial manufacturer who convenes a review board strictly for warranty considerations. Government contractors, however, find it an extremely convenient tool to use for communication between themselves and their customers. Most large contractors include the warranty review board meetings in conjunction with other quarterly or biannual reviews, such as the integrated logistics support meeting, and the annual cost review. The value of having the warranty reviewed periodically cannot be overemphasized. It provides clarity on disputed issues, keeps a running record of events, and generally leads to a much closer relationship between contractor and government individuals working in the same field. This author highly recommends quarterly or semiannual reviews between the buyers and sellers, because experience has shown that most of the problems caused on other programs without review boards were caused due to lack of communication. Face-to-face discussions seem to resolve these problems more effectively.

Lessons Learned

Customer-supplier relationships

The past few years have shown a marked change in the commercial market in regard to warranties. Previously, many sellers had a very weak, if any, warranty. Only a few leaders in the field, such as Curtis Mathis, understood the power and effectiveness of featuring a warranty as one of the key selling points of their product. There is no longer any doubt that a warranty which is superior to the competition can be one of the main selling features of a variety of products, from cars to washing machines. Lee Iacocca has brought Chrysler back into the competitive market with his 7-year, 70,000-mi warranty. Even more creative warranties are now being offered, including having an option to pick different warranties on the same product. The buyer now expects a good warranty on a product, and the sellers have responded to this demand. However, the author still feels that most warranties are handled as an afterthought by many producers, and investigation into many companies has shown that a cost analysis seldom is done to determine whether a warranty will produce a profit or loss. The marketing and sales departments are the ones pushing the manufacturers to offer the new, improved warranties, and no actual analysis is performed after sales to determine the impact of the warranty, beyond increased sales. This author has been successful in negotiating for additional terms of a warranty with an automobile dealer. Even for commercial warranties, negotiating works!

Continuous negotiation

In the government warranty field, most people feel that once a contract is received and signed, the negotiation process is complete. Far from it! The real negotiation process has just begun. It occurs in the implementation of the warranty, and in the interpretation of the requirements. Often during a production contract, it would be advantageous to the customer and the producer to make modifications to the existing warranty. It is the warranty manager's main task to analyze the production warranty to see if changes would produce a more workable warranty by implementing those changes which are acceptable to the customer. If it can be shown to the customer that a change will be of benefit to both parties, it usually meets with little resistance. It is the warranty manager's job to find and show the benefits to the customer. No one is an expert in the warranty field, and the customer is glad to respond to recommended changes when it will benefit the program. In many cases it may even reduce cost; all the better. Government agencies are always looking for ways to reduce costs, and those suggestions are readily accepted. Lost revenue to the seller is more than offset by the customer satisfaction and future business gained by such action.

Importance of review boards

The formation of review boards consisting of parties from both sides of the contract have proven on existing production programs to be one of the most important aspects of the warranty program. Meetings held periodically, it is recommended that reviews be held at least quarterly for the first year or two of a new program, allow the players from government and industry to meet and discuss the status of the warranty, make suggestions for changes, review performance, and build a warranty team that works toward the common goal of customer and equipment satisfaction.

It is also a good idea to set up periodic meetings between the prime contractor and the subcontractors to review the results of the subcontractors' warranty procedures, configuration data, and testing results. It has been this author's experience that few subcontractors keep adequate records of testing or configuration, and if a recall or retrofit is necessary because of systemic problems of the product, it is difficult to show liability without proper documentation. Mutual trust between the buyer, seller, and subcontractors is paramount, but good documentation is good business.

Importance of cost tracking

The importance of cost tracking cannot be overemphasized! The customer must be assured that the warranty is cost effective, and there is no way of knowing that without knowing the total cost. A cost-effectiveness analysis is to be performed prior to contract award, but in many cases this has not been accomplished. Most government agencies must report the cost of a warranty at least annually, and trying to get the figures after the fact is almost impossible. If quarterly or semiannual reviews are held, this is a good time to review and update costs.

The importance of cost tracking to a contractor can be the difference between a profit or a loss. Even though the customer may not be billed for warranty repairs, the company management needs to know how much this activity is costing, so that decisions can be made in regard to warranty repair policy. If retrofits or recalls are necessary, some of the cost may be passed on to the subcontractors if it is documented accurately and in a timely manner. Recall activity sometimes lingers on for years, and a properly executed accounting procedure for accumulating costs is a must. The author was involved in litigation with a government agency on a major production program which had a retrofit-recall program that ran into the millions of dollars. Much of that was recovered from subcontractors, after adequately segregating that cost from other warranty activity. As mentioned earlier, a warranty can be profitable to the seller, while being acceptable to the customer, but only if the warranty procedures are adequate to effectively administer the warranty program.

Chapter 11

Warranty Development

General Considerations

Up to this point we have presented the key principles of warranties, including the concepts, tools, and procedures. In this chapter we will discuss the development activities necessary to help ensure the thorough and timely integration and application of these principles. Our goal continues to be a win-win situation for both the supplier and the customer. We will discuss the development activities relative to the customer first, and then the supplier. Because of its more specific requirements, the majority of the discussion will be relative to the government business sector.

Government Development Activities

Warranty program objectives

Discussion of government warranty development activities must begin with a clear understanding of the warranty program objectives. These objectives are to

- Develop and acquire warranties that
    - Motivate the contractor to ensure product quality and performance.
    - Continue contractor responsibility and involvement beyond the delivery date and for the entire warranty period.
    - Are easy to manage and administer, such that there is no disruption to existing military systems and procedures.

  - Are enforceable.
  - Are affordable in relation to potential benefits.

- Provide standard procedures for identifying, reporting, tracking, and correcting defects and failures covered by a contractual warranty, including performance measurement and tracking of weapon systems, equipment, and items.
- Minimize the need for new and costly warranty data tracking systems and related workforce resources to administer contract warranties.

It is essential that every one of the objectives be considered in the development of the warranty clause. Failure to properly address any of these factors could jeopardize the cost effectiveness of the implemented warranty.

Warranty and the system life cycle

The following is a discussion of warranty-related activities from a system life-cycle perspective. To develop an effective warranty, the program manager must plan for the execution of these activities. Also addressed are warranty impacts on the acquisition strategy and procurement plan, the system specification, and the program office organization as key planning factors for the program manager to consider early in the system's life cycle. Contractor risks are also discussed.

Life-cycle overview. The reader is referred back to Fig. 2.6, which shows how warranty-related activities interface with the system life cycle. These activities are summarized by phase as follows:

- *Concept exploration and definition.* Technical and support concept studies are performed to identify characteristics for warranty consideration.
- *Demonstration and validation.* The expected warranty provisions are developed as system requirements.
- *Engineering and manufacturing development* (*EMD*). The warranty provisions from demonstration and validation are updated to reflect better estimates of system reliability and maintainability, support parameters, and costs. The provisions are then incorporated into the production RFP. A series of tasks to implement, enforce, and manage the warranty provisions is developed and coordinated.
- *Production and deployment.* Tasks to implement, enforce, and manage the warranty provisions are finalized.

- *Operation and support.* The warranty provisions are implemented and administered.

Acquisition strategy. To obtain maximum effectiveness, it is important that the warranty concept be considered early in the weapon system's life cycle. Decisions on equipment configuration and design affect the warranty approach as well as the planning needed to maintain and support the warranted system.

The following is a general sequence of steps to develop a warranty strategy, starting early in the system's life cycle:

- Perform studies to identify essential performance characteristics to consider for warranty and identify candidate approaches.
- Develop criteria and models and collect applicable data to perform evaluations to decide between assurance and incentive types of warranty.
- With technical, user, logistics, and contract personnel, develop candidate approaches and assess the feasibility of candidate approaches, including implementation and administration.
- Develop preliminary clauses or draft provisions for demonstration and validation RFP.
- Issue EMD RFP with expected warranty provisions for the production contract, or have the contractor propose alternative forms of warranty to the government.
- Finalize warranty terms and conditions for the production RFP.
- Develop a warranty selection strategy and decision model.
- Issue an RFP for production with a warranty option.

System specification. A key element in the development of an effective warranty is the system specification. It defines system requirements. Ordinarily, it is developed prior to the completion of the demonstration-validation phase. Specific recommendations to include as requirements in the specification, considering warranty development, are

- Requirements in the system specification and flow-down specifications must be quantitative.
- Requirements used directly for warranty coverage must clearly define the operational or special test conditions.
- Methods to determine conformance to requirements must exist or be amenable to development.

- Only a small subset of specification requirements should be selected for warranty coverage.
- Higher-level, mission-related requirements are generally preferred to sublevel requirements for warranty specification, for example, system MTBF instead of subassembly MTBFs.

Program coordination. It is the program manager's responsibility to plan, coordinate, and integrate warranty application as early in system development as possible. The selected warranty approach should serve as a lever to enhance system reliability by configuration, design, and maintenance and support parameters. Essential performance warranties should be fully integrated into the weapon system program.

The program manager is responsible for ensuring that the system warranty is developed and implemented effectively. The military services and program managers should designate a warranty manager to act as the focal point for warranty task performance. The warranty manager serves as the functional interface between the program manager, user, contracting officer, and support activities.

The warranty must be consistent and compatible with operational and logistic concepts and with the overall acquisition strategy. To secure consistency and compatibility, the team concept should be employed from the outset. The program manager should involve all using and supporting commands and agencies throughout the planning process. Functional interfaces between the program office, user, and supporting activities ensure maximum benefit from warranty application.

Contractor risk considerations. New procurements involve significant technical, operational, schedule, and financial challenges. A warranty is a means to shift part of the development and acquisition risks to the contractor. However, if consideration is not given to the risks the contractor assumes when undertaking a warranty, the effectiveness of the warranty is undermined. Warranty price will increase as uncertainty increases. It is unreasonable to ask contractors to incur extraordinary financial losses, the reasons for which were not reasonably foreseeable. The viability of the entire program might be threatened.

Concept exploration-definition phase activities

The program manager evaluates and selects alternative system development concepts to meet the stated mission need. The concepts should address the functional and performance characteristics necessary to meet the mission need along with any necessary interfacing capabilities. They should be accompanied by preliminary life-cycle cost (LCC) estimates and logistics supportability plans.

Although the system is treated in general terms, evaluations may be conducted in terms of system reliability and projected LCCs. Warranty or other control methods (e.g., award fee) may be considered means to achieve stated goals for reliable performance pursuant to 10 USC 2403 (U.S. Code Title 10, Sec. 2403) and maintain costs within resource limitations. Program documentation should clearly reflect initial criteria to employ warranty control techniques.

Demonstration-validation phase activities

The program manager identifies the system development concepts and approaches that have the greatest potential to meet the mission need in the most cost-effective manner. The concepts are verified and associated risks and uncertainties are identified and resolved where possible, usually through trade studies, models, prototypes, and demonstrations. System and subsystem documents as well as solicitation documents are completed to support contracting for the EMD of the selected concepts.

Although warranty application is generally associated with the production contract, warranty requirements may influence design, production processes, parts selection, and quality control in an effort to enhance reliable system performance. The RFP for EMD should contain preliminary warranty provisions intended for use in the production contract.

The program manager must determine a warranty approach to the weapon system and identify preliminary terms and conditions for the warranty. A structured approach to warranty development is a step by step process as follows:

- *Initial screening.* Initial screening is performed to determine whether one or more warranty alternatives are appropriate.
- *Economic analysis.* If the initial screening results are positive, the candidate warranty alternatives are analyzed to determine the economic implications and appropriate warranty period.
- *Development of provisions.* Initial warranty provisions are developed. The program manager should maintain continuous coordination with using commands and support activities.
- *Incorporation of provisions in EMD RFP.* After proper initial review with cognizant procurement, legal, and other pertinent agencies, the initial warranty provisions are incorporated into the EMD RFP—primarily for informational purposes, unless a firm or not-to-exceed

warranty commitment must be made at this time. Special bidder instructions may be necessary to clarify selected points. Additionally, special briefings with potential offerors may be necessary to elaborate on the intent of the provisions.

- *Development of final preliminary provisions.* As a result of the foregoing processes, the initial provisions may be developed to clarify wording, changes in coverage, and other issues. In the case of a combined EMD and production-deployment procurement, the final provisions may become part of the contract, typically as an option that may be exercised at a later point in EMD. If it is not a combined procurement, the provision may still undergo additional changes and evaluation as part of the production procurement.

EMD phase activities

The EMD phase culminates in a baseline configuration design. It also results in a documentation package that reflects the established cost, schedule, logistics supportability, and performance constraints.

During the EMD phase, better estimates of system reliability, maintenance and support parameters, and operating capabilities generally emerge. Warranty applicability and economic studies should be refined and updated, and warranty provisions should also be updated to reflect program or equipment modifications that have occurred during this phase. Major warranty evolutions in this phase are summarized as follows:

- *Warranty feasibility assessment.* The initial economic analysis performed as part of the demonstration-validation phase should be updated or refined in light of current information. If a previous evaluation was not performed, an assessment should be initiated.
- *Development of final provisions.* If warranty provisions were not finalized as part of the demonstration-validation phase, provisions for the production-deployment phase are formulated or refined, with thorough coordination between program manager, support activities, and users.
- *Production RFP provisions.* Provisions are incorporated into the production RFP if they were not incorporated previously. Warranty issues addressed in the RFP include warranty management, claim processing, dispute procedures, facilities and equipment, in-plant material flow, warranty data, price, and prior performance. Instructions to bidders regarding required responses may be necessary.
- *Proposal review.* Production proposals must be evaluated with respect to the warranty response. The spirit and intent with which those offering the proposal address warranty provisions as well as

quoted price are the prime concern. If a warranty price quotation is obtained, the economic analysis should be reperformed using the quoted warranty cost in lieu of estimates. Any questionable points may be clarified in discussions held with contractors. The reader is directed to the *Warranty Guidebook* (1992) published by the Defense System Management College, page 5-11, Table 5-6 for an excellent discussion of proposal evaluation factors.

- *Warranty decisions.* On the basis of the economic analysis, as well as mission and logistics factors, the program manager must decide among available warranty options. The decision should be made early enough to permit orderly planning by all affected activities, regardless of the choice made. If a warranty is selected, provisions to fund and effect warranty payments must be established.

Warranty selection factors

The following subsections address factors related to acquisition, the system, and operation that can influence warranty selection and warranty terms and conditions.

Acquisition factors

- *Development history.* Detailed data available on the system should be used to determine potential warranty problems. Prediction and test data can help define quantitative warranty requirements.
- *Small versus large buy.* The larger the buy, the greater the potential risk to the contractor if warranty terms and conditions are not met. For a small buy of large, expensive items, the warranty duration may be administered on an item-by-item basis. For a large-quantity buy, trying to manage warranty duration on an itemwise basis may evolve into extensive administrative problems. Accordingly, warranty duration on a population basis, such as a single end date for all systems is recommended.
- *State of the art.* The greater the technological challenge, the more difficult it is to structure a fair warranty at an equitable price. However, it is the technological challenges which must merit warranty consideration. EPRs of weapon systems that "push" the state of the art are prime candidates for warranty.
- *Competition.* The degree of competition affects warranty price and contractor's enthusiasm to undertake or bid warranties with some risk. Without competition, it is generally better to impose warranty requirements rather than have a sole-source contractor bid. Warranty terms and conditions should not inhibit plans to compete future production contracts.

System characteristics

- *Electronic versus mechanical.* Many electronic systems exhibit relatively constant failure rates over time, which makes warranty duration a less important factor than for mechanical systems which wear out in proportion to time and use. Given a limited historical database, there will generally be more confidence in a warranty analysis of electronic systems than in an analysis of mechanical systems. Electromechanical characteristics are therefore important considerations for warranty duration and reliability prediction.
- *Transportability.* Should the warranty be so structured as to require contractor repair, the ability to economically and expediently ship failed systems and components requires consideration. For warranted systems that are excessively large, a warranty remedy requiring contractor in-plant repair is not feasible. "Ruggedization" is another factor in developing warranty terms and conditions that requires transporting systems to another facility.
- *Field testability.* The ability to reliably determine whether a weapon system has failed is important for maintenance concept and warranty integration. If adequate equipment or procedures are not available to test weapon systems, then a significant number may be sent to the contractor for warranty action which, in fact, exhibit no evidence of failure. This can be costly to the government if the contractor can charge for processing nonfailed systems.
- *Warranty markings and seals.* Ideally, warranted weapon systems and components should be clearly marked with appropriate warranty data and instructions. Markings on shipping crates and boxes contribute little. If a weapon system or component cannot be suitably marked, or if it cannot be protected against unauthorized maintenance through seals or other controls, warranty terms and conditions should be appropriately adjusted. The government should, in any event, retain the discretionary right to break any seal to take necessary corrective action.

Operational factors

- *Acceptance-employment cycle.* The length of time from weapon system acceptance until placement into service should be factored into the warranty duration. Acceptance typically occurs at the contractor facility with the signing of a DD Form 250 by a government representative. Placement into service may occur months, possibly years, later as weapon systems wind their way through shipping, inventory storage, and distribution cycles en route to actual employment. Either the average "transit" period can be added to the length of the warranty, or the warranty can be definitized to commence on employment rather than acceptance.

- *Operating cycle.* Weapon systems may be employed only once, for instance, a missile, intermittently, such as an aircraft, or continuously, such as a surveillance radar. Such types of usage affect reliability performance parameters as well as measurement criteria for success or failure in field use. For one-shot usage, probability of success is the most applicable reliability parameter; for intermittent usage, MTBF may be more appropriate; and for a continuously operating system, operational availability could be the most critical consideration.
- *Existing military maintenance capability.* If military maintenance capability already exists, a warranty necessitating a contractor to establish a repair facility is not likely to be cost effective. This does not rule out alternative forms of remedies that do not require contractor repair facilities.
- *Performance measurement.* The ability to measure performance parameters is critical in establishing EPRs. While elapsed-time indicators and meters on weapon systems may be used to record operational usage, maintenance records may also be used to record failures as well as process warranty claims. In some cases, special data-collection methods may have to be employed or special operational tests conducted. As a general guideline, the probable success of a weapon system warranty program is inversely proportional to the efforts involved in determining whether a breach has occurred.
- *Pipeline factors.* The transportability of the weapon system, length of the pipeline, sparing level, and cost of spares all influence the maintenance concept of the weapon system under warranty. Government repair using bill-back procedures may be a viable option should contractor repair prove unsuitable because of pipeline factors.
- *Self-sufficiency.* Warranty remedies using bill-back procedures may also be appropriate in cases where criticality or performance dictates military maintenance.
- *Transition.* Termination of the warranty can have momentous consequences. Considerable thought should be given to a one-time transition over a phased transition, especially if the contractor performs depot-level maintenance.

Warranty alternatives

The following sections identify alternatives for consideration when structuring a warranty, starting with the basic need for a warranty.

Warranty versus no warranty. Figure 11.1 depicts the process that should be applied to determine whether a warranty is required pursuant to Title 10, U.S. Code Sec. 2403.

EPRs versus no EPRs. Figure 11.2 shows the process to be used in determining the need for and application of essential performance requirements (EPRs) in the warranty.

Assurance versus incentive warranties. As discussed in Chap. 2, the assurance warranty assures the government that minimum performance and quality requirements are satisfied. An incentive warranty helps motivate the contractor to exceed contractual requirements. The degrees of coverage and commitment separate the assurance warranty from the incentive warranty. Figure 11.3 presents a decision diagram to assist in choosing between assurance and incentive warranties. Although developed for Navy applications, the diagram has broader application. The diagram is based on the premise that an incentive form of warranty is most applicable when certain conditions exist. These conditions are discussed in Chap. 2. As was also pointed out in Chap. 2, not all these conditions may exist for any one program, and their absence does not preclude the use of incentive warranties. In the diagram, factors denoted by *D* (dollars), *P* (period), *M* (missile or ordinance), *S* (ship, ship system; satellite), and *R* (repair by contractor) indicate when certain conditions are violated.

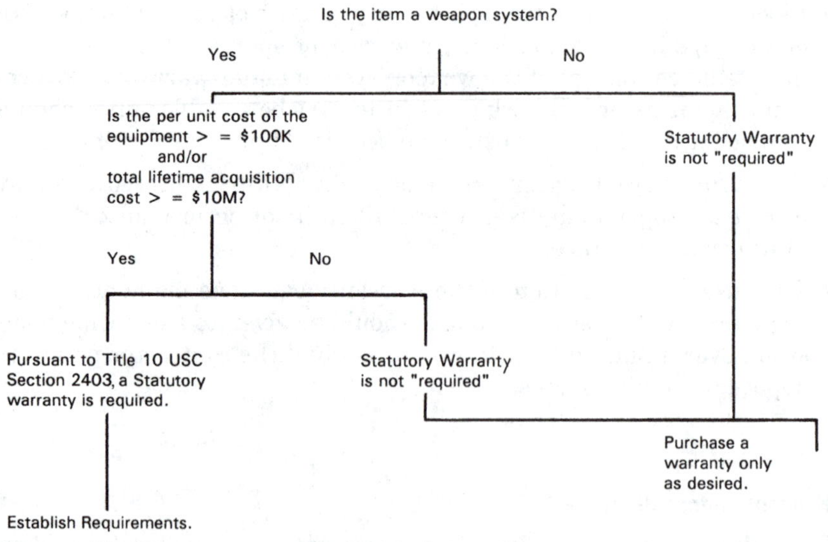

Figure 11.1 Determination of warranty need.

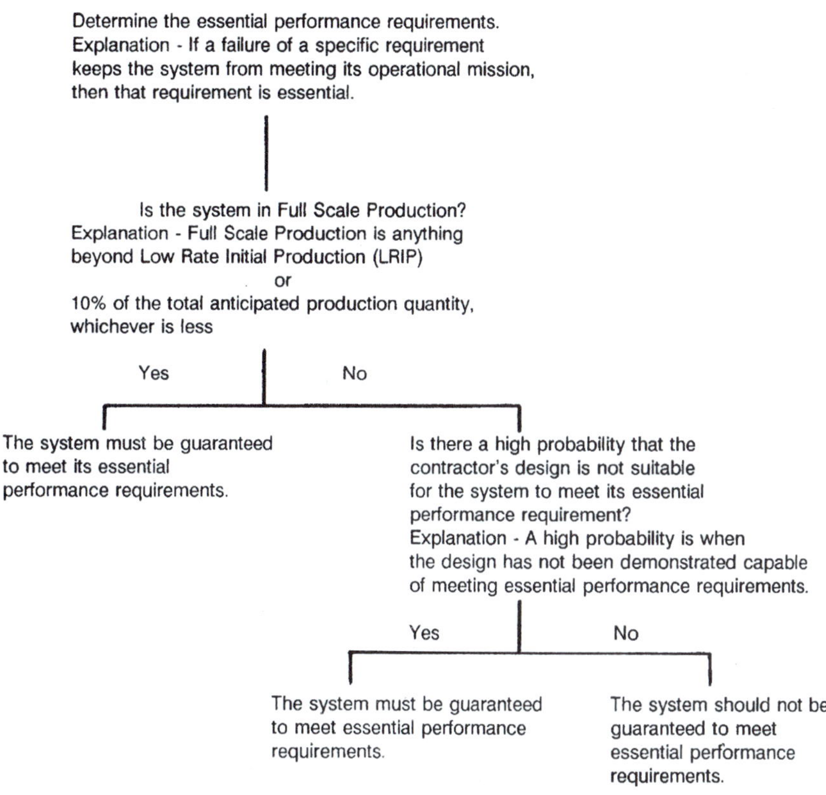

Figure 11.2 Determination of essential performance requirement need.

Individual versus population controls. A warranty can be placed on an individual system, the population of systems, or both. Normally, the warranty coverage related to defects in materials and workmanship applies to individual systems. Coverage of design and manufacturing requirements and also of EPRs may apply to either the individual system or the population. Design problems are related to the entire population.

Relative to EPRs, in terms of controlling reliability, an MTBF guarantee usually applies to a population of weapon systems or equipment. However, it is possible to apply such a guarantee to individual weapons. For example, a contractor may supply several communications satellites and provide guarantees as to the number of communication channels available on each individual satellite.

The type, quantity, and cost of the warranted weapon systems will often dictate whether population or individual system coverage is preferable. Large buys of small systems, such as an avionics system, are often more appropriate for population coverage, while small buys of large systems, such as command, control, and communications systems, are better suited to individual system coverage.

Special tests versus operational performance monitoring. Means to determine conformance of actual weapon system performance to EPRs must be identified. Two approaches are

- *Special operational testing.* The contract specifies a test to measure one or more parameters to determine conformance to the EPRs.
- *Operational performance monitoring.* Data is collected during normal operations and used to calculate relevant operational statistics for comparison to the EPRs.

Use of special test procedures allow for direct and accurate measurement of characteristics of interest. However, because of the high cost, the tests are generally short in duration and the data obtained may not produce a high level of confidence. Performance monitored during nor-

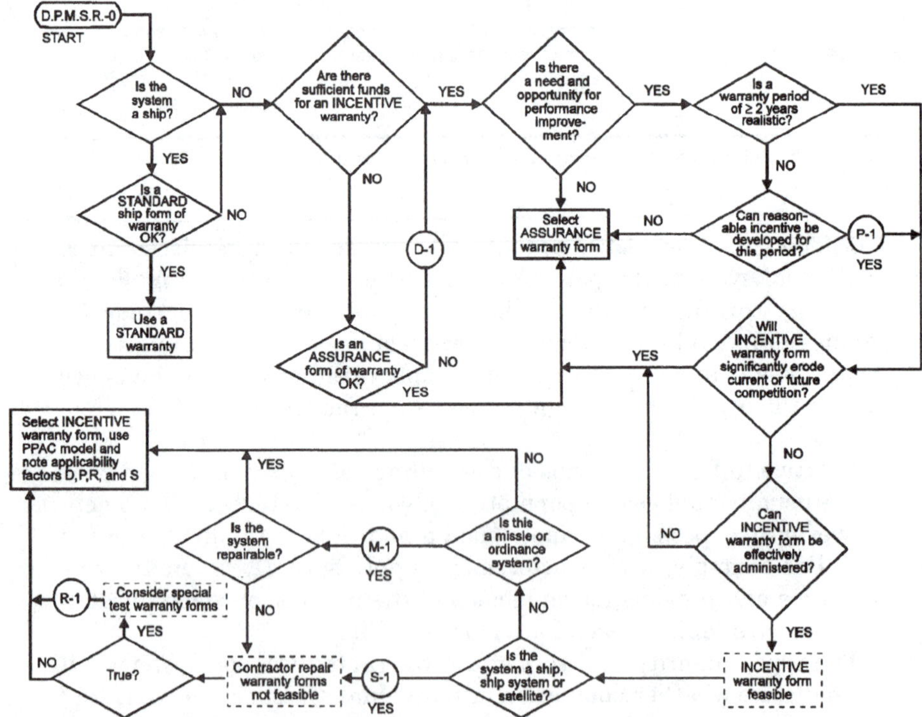

Figure 11.3 Assurance versus incentive warranty decision.

mal usage allows for a much greater sample size, but is invariably labor-intensive, expensive, and highly subject to measurement and recording errors.

Failure-free versus threshold warranties. Figure 11.4 presents the process that should be applied in determining the proper repair warranty type to be used on a given acquisition. The reader is referred to Chap. 2 for the definitions of failure-free and threshold warranty types.

Individual-item versus systemic coverage. Once the appropriate warranty type has been selected, it is then necessary to select the appropriate coverage for the warranted system and its components. Figure 11.5 depicts the process to be applied in determining the proper warranty coverage to be invoked on a particular warranty. The reader is referred to Chap. 2 for definitions of individual-item and systemic coverages.

The warranty plan

Warranty plans should be developed by the program office and coordinated with the using and supporting organizations, as well as the cog-

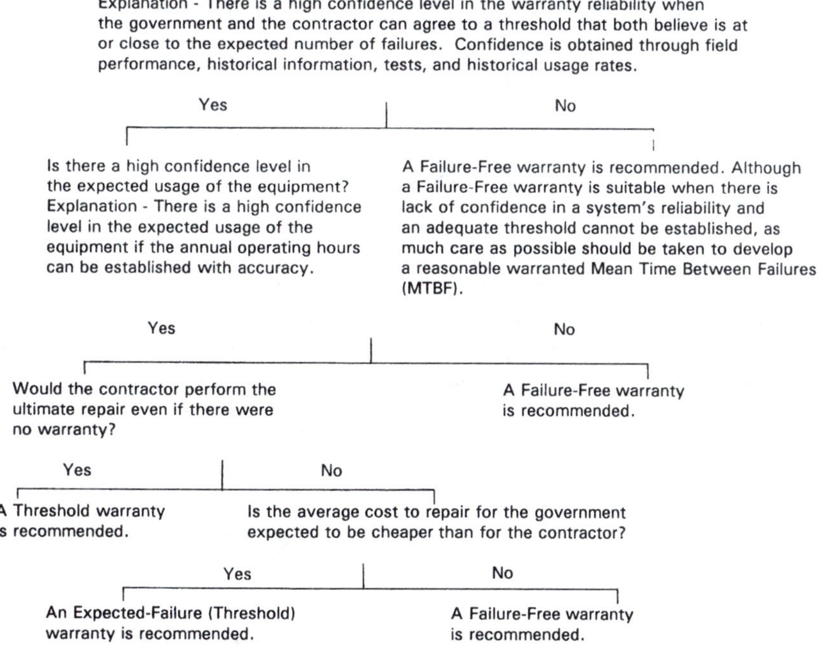

Figure 11.4 Selection of repair warranty type.

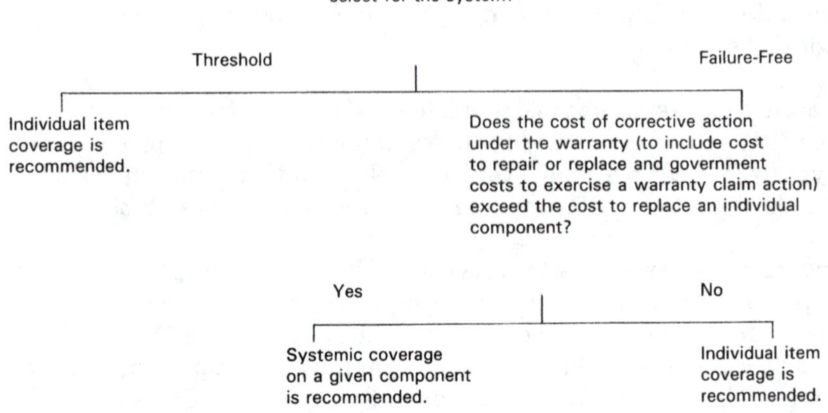

Figure 11.5 Selection of warranty coverage.

nizant contract administration office, and any other organizations tasked in the plan of warranty support. The program manager has overall responsibility for warranty planning and the establishment of the warranty team to prepare and coordinate the plan.

The warranty plan should be approved by the program manager within 6 months after the EMD contract award, and updated as needed to provide warranty implementation requirements for fielding the warranted item. The warranty plan should also be updated to reflect any change in requirements prior to the award of follow-on production contracts.

The warranty plan should address the following:

- *Acquisition background.* Describe the system being acquired. Summarize the program and warranty history to date.
- *Warranty clause.* Include the proposed warranty clause with the plan and identify any special considerations or constraints affecting selection of the terms and conditions.
- *Cost-benefit analysis* (*CBA*). Describe the CBA methodology to be used, and summarize any CBA results to date.
- *Warranty administration.* Describe the specific requirements to administer the warranty. Ensure that the administrative requirements of the proposed warranty clause are consistent with this section of the plan.
- *Warranty team membership.* Describe the warranty team organizational and management responsibilities. List the team membership: warranty manager, contracting officers, engineers, logisticians, cost

analysts, using and supporting organization representatives, and other points of contact deemed necessary for warranty administration.

- *Contractor support.* When contractor support, contractor logistic support (CLS), or interim contractor support (ICS) is planned, ensure that the support requirements are clearly defined, compatible with the warranty, and the related costs of each (warranty and CLS or ICS) are segregated for accounting purposes.
- *Schedule.* Identify key events and dates such as delivery dates, warranty periods, and CBA accomplishment and updates.
- *Preparation of the warranty clause.* Using the guidance of Chaps. 3 and 6 and the warranty plan, we are prepared to construct the warranty clause.

The terms and conditions of the warranty must be tailored to the system under study and must be as clear and simple as possible with emphasis on enforcement of the warranty conditions through the existing management, administration, and logistics processes.

The following requirements should be included in the warranty terms and conditions unless the warranty plan provides rationale for the exclusion of the requirement and approval has been granted, if required:

- Define key terms such as acceptance, defect, remedy, etc.
- Incorporate the three guarantees required by Title 10, U.S. Code Sec. 2403.
- Describe the roles and responsibilities of the government and contractor in the warranty process.
- Identify the production units covered by each of the three parts of the warranty and the units, if any, excluded from the warranty coverage.
- To the extent possible, state the warranty duration as a fixed period of time from date of delivery. The duration must be of sufficient length to determine that all the warranty requirements have been achieved. Warranty duration should allow for those anticipated nonoperational activities after delivery such as transportation, storage or shelf life, and redistribution. When the duration is based on item utilization rather than calendar time, appropriate measuring devices (elapsed-time indicator, cycle counter) and procedures are required. Individual-item coverage duration should be on a per unit basis, while systemic coverage should cover a period of calendar time from acceptance of the first warranted unit to a reasonable time beyond acceptance of the last warranted unit. The warranty period

could be extended under certain conditions. i.e., lengthy repair or replacement times.

- Describe the EPRs to be warranted. Include a description of how they are to be measured, when they are to be verified, and any special testing or test equipment required to complete the verification.
- Clearly state the marking requirements in terms of the information to be conveyed and the proper location on the warranted unit.
- Clearly describe the warranty coverage.
- Describe the remedies available to the government if the contractor breaches the warranty. Conditions for invoking a particular remedy should be addressed, including whether redesign is a remedy and under what circumstances it would be invoked.
- Describe all warranty data and report requirements and include appropriate contract data requirements list (CDRL) for distribution to the appropriate organizations.
- Identify any exclusions such as fire, flood, combat damage, and mishandling.
- Identify any limitations such as contractor's financial liability.
- Establish warranty terms and conditions consistent with the system's operational and maintenance concept.
- Include a statement that the warranty does not limit the government's rights under any other clause.
- Establish packaging and handling requirements for warranted items.
- Establish transportation requirements for the government and the contractor, including cost distribution.
- Address the prime contractor's responsibilities for warranting government-furnished equipment (GFE).

As was pointed out in Chap. 3, the mutual success of the warranty is directly proportional to the clarity and thoroughness of the warranty clause.

The reader is encouraged to evaluate the warranty clause presented in Chap. 3 against the suggested content just presented.

Contractor Development Activities

Warranty team

Figure 7.3 presents the structure of the contractor warranty team. Just as a well-informed, integrated team was an essential ingredient in the preparation of clear and thorough warranty requirements, the team concept is equally important relative to the contractor's response to these requirements.

All team members bring knowledge and experience in their respective disciplines along with familiarization of the system to be warranted. The thoroughness and reasonableness of the contractor's warranty response is directly proportional to the experience and cooperation of the team members.

The cost analyst, who is the catalyst of the team, should

- Formulate the warranty cost-risk model for warranty costing and warranty tradeoff studies.
- Define the cost-risk model input parameters for each team member, as applicable, to estimate.
- Perform all warranty costing and tradeoff analyses.
- Interpret results of warranty costing and tradeoff analyses to the program manager and team members to aid in the decision-making process.
- Support warranty negotiations with the government.

The program manager, who is the executive officer of the team, should

- Staff the team.
- Set the schedule for warranty milestones, culminating in the warranty quote to the government and subsequent negotiations.
- Ensure the necessary integration among the team members.
- Ensure the timely estimation of required warranty analysis input data from the engineering members of the team.
- Make the final decision on warranty tradeoff analyses.
- Set the warranty price for the quotation to the government.

The engineering members of the warranty team are the foundation of the warranty analysis efforts. These team members should

- Define warranty tradeoffs in their respective disciplines.
- Estimate required input data in their respective disciplines for warranty tradeoff and costing analyses, including supporting rationale.
- Assist in the discussion of tradeoff and costing analysis results leading to the final decisions.

The contract specialist has important administrative responsibilities relative to the warranty process and should

- Determine whether similar warranties have been quoted, and, if so, notify the team members of the details.
- Ensure the warranty response reflects the findings of the tradeoff and costing analyses.
- Ensure the warranty response is properly stated to minimize contractor risks and legal red tape.
- Coordinate the evaluation and wording relative to alternative warranty offerings, as appropriate.
- Lead in warranty negotiations with the government.
- Update the warranty history database.

Warranty planning

There are important contractor warranty planning activities leading up to the warranty quotation to the government. These activities include the following:

- Researching the customer
- Researching the competition
- Structuring the warranty team
- Training the warranty team
- Preparing the warranty statements for candidate options
- Evaluating the cost-risk of candidate options
- Choosing the "best value" warranty option
- Setting warranty price to the government
- Documenting thoroughly the warranty quote and supporting rationale in preparation for negotiations with the government

In researching the customer, it would be helpful to be familiar with warranty requirements for other programs managed by a particular program office. It is possible that these requirements might be restated in the current warranty clause. Also, it would be helpful for the contractor to be aware of the customer's budget for the warranty, or, at least, for the total program. This data would help to more intelligently set the warranty price and therefore determine the contractor's risk in quoting that price.

Knowledge of the competitor(s) is useful. Competitor(s) warranty performance history could provide insight into areas that could be emphasized in the contractor's current warranty response to capitalize on the weaknesses or offset the strengths of the competitor.

The importance of the identification and functioning of the warranty team has been discussed previously. Without the coordinated team effort, the effectiveness of the entire warranty planning effort is severely hampered.

Once the warranty team has been staffed, it is essential that they be thoroughly trained. The cost analyst is an excellent candidate for conducting the training. The team should receive training in the specific requirements of the current warranty, with emphasis on any penalties or incentives, or unusual remedies. The team should also be familiarized with the quoting process, with emphasis on their particular role in the cost-risk analysis process.

Assuming the contractor is offering an alternative warranty to the requirement in the RFP, it is important that the alternative warranty statement(s) be prepared in a way to facilitate their evaluation. They should be clearly and thoroughly stated, emphasizing the variances from the RFP requirement.

The evaluation of the cost and risk associated with any alternative warranty options is crucial to the selection of the best alternative option. For example, for a specific warranty price, the risk incurred by the contractor can be determined for each option. Also, for a specific chosen contractor risk, the corresponding warranty price can be determined for each option.

The choice of the best-value warranty option is based on factors such as

- Coverage for the government.
- Cost to the government.
- Profit potential for the contractor.
- Competition.
- Risk for the contractor.

These factors may be traded off using cost-risk analysis to arrive at the best mix and therefore define a new alternative option.

Once any tradeoff analyses have been performed, the warranty price can be set for any alternative option(s) as well as for the RFP requirement. In setting the warranty price(s), ample use is made of the results of the cost-risk analysis to determine a competitive warranty price within an acceptable risk level.

Not enough can be said regarding the importance of the documentation of supporting rationale for the warranty quote. Negotiations on the warranty quote could come as many as 3 years after the original quote. It is virtually impossible to reconstruct the details of the warranty quote, including the assumptions, the model, the calculations, and the results, after that time lapse. Without detailed rationale, the negotiating efforts are severely hampered, and could result in added risk to the contractor.

As an exercise, the reader is encouraged to evaluate the warranty example at the conclusion of Chap. 5 to determine whether this alternative warranty offering could be considered a "winner" by you, the customer. In performing this evaluation, consider the following situation:

- You are the missile prime contractor.
- You must offer an alternative warranty per the statute in addition to the RFP requirement.
- Your company says this is a "must win" program.
- This is the EMD phase proposal.
- The customer says thoroughness of the warranty offering is a key evaluation factor.
- Competition is stiff.

In making your evaluation, be sure that this alternative warranty satisfies the warranty statute. Would you choose this warranty? Why? Why not? The solution is left to the reader.

Manager's warranty checklist

As a summary of contractor development activities, the following checklist used by the program manager is provided:

- Understand and quote RFP requirements.
- Propose alternative warranty, as applicable.
- Use all available history and/or quote data.
- Use team approach for quoting warranty.
- Perform bottom-up warranty analysis, including cost-risk analysis (no generalized percentages).
- Address administration, field service, repair, redesign or retrofit, transportation, and performance parameter penalty costs, as minimum, in the cost-risk model.
- Document supporting rationale for the quote.
- Prepare thoroughly for negotiations, including knowing your customer.
- Establish thresholds on your vital quote parameters.
- Demonstrate patience in the negotiations.

- Realize that effective negotiations take time and lots of homework.
- Establish a partnership with the government negotiator(s) rather than being an adversary.
- Consider warranty as an opportunity for profit and customer satisfaction.

The last item on the checklist is the most important since it essentially drives the others. Contractors should seek to make a profit on all their business ventures, warranty included. Not enough can be said regarding the importance of the warranty relative to overall customer satisfaction. Chapter 12 expands on the issue of customer satisfaction relative to warranties.

Joint Army-Industry Warranty Working Group (JAIWWG)

As an example of current warranty developmental efforts within the Department of Defense, the work of the JAIWWG is presented. The author was privileged to participate in the JAIWWG effort as the industry cochair for the Warranty Requirements Determination working group. The effort was prompted by the realization that the Army, in some cases, was procuring warranties that did not make good business sense in terms of their cost effectiveness.

Mission

Develop Army warranty requirements, evaluation, and administration that meet the intent and the requirements of the statute and make good business sense.

Charter

- Review existing Army warranty processes.
- Recommend improvements to existing processes in the areas of requirements determination, proposal or negotiation evaluation, and administrative procedures.
- Achieve timely and effective Army-industry interaction in identifying and resolving key warranty issues.
- Structure and process. Figure 11.6 presents the JAIWWG structure and process.

As Fig. 11.6 shows, the requirements, evaluation, and administration working groups provide inputs to the implementation working

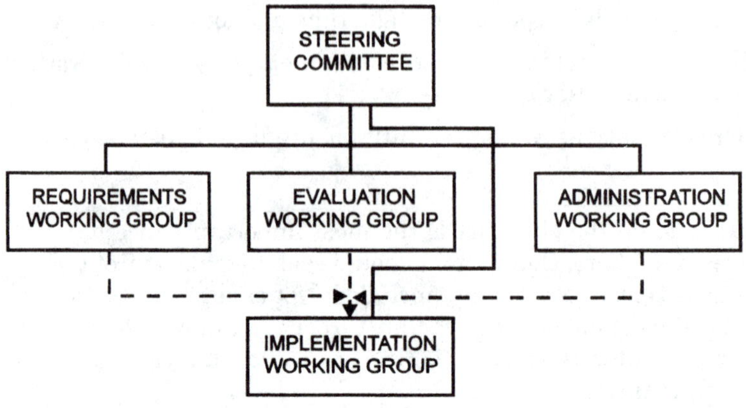

Figure 11.6 Joint Army-Industry Warranty Working Group (JAIWWG) structure and process.

group to enable the generation of concrete implementation activities. All four working groups take direction from the steering committee.

Membership

All five major Army Commands and Army Materiel Command (AMC) were represented. The industry companies included Texas Instruments, AAI, Magnavox, Westinghouse, Rockwell, GTE, Harris, and McDonnell-Douglas.

Findings

The current Army warranty process was found to be inadequate. The requirements are inconsistent and incomplete. The evaluation criteria are nonexistent or too vague. The administrative procedures lack consistency and incentive.

Actions

The structure, wording, and process for determining warranty requirements were defined. Cost-effectiveness analysis was introduced as the evaluation tool. Procedures for warranty item processing were defined.

Army-industry interaction

There was some concern at the outset that industry concerns would be heard, but through spirited exchanges between the Army and industry and an overall cooperative atmosphere, both now have a better understanding of key issues, and as a result both Army and industry were heard.

Implementation challenges

The two major challenges to implementation are the institutionalized training of the Army at all levels and contractors, and the revision of existing Army warranty documentation to include regulations, pamphlets, and similar. Both implementation vehicles must be accomplished in a timely manner.

It is hoped that the JAIWWG recommendations will have application within the Navy and Air Force, and that these services can benefit from the JAIWWG activities to preclude duplication of effort. The Army should be commended on their willingness to admit their shortcomings relative to warranties and to take positive corrective actions to correct these deficiencies, including industry participation.

Chapter

12

The Future of Warranties

Global Economic Outlook

Historically, the United States has produced products that were far superior to those of other countries of the world. In essence, we had a corner on the market. In recent years the Japanese and others have drastically improved their products and have become dominant in many markets. The Japanese dominance is most evident in the automotive and electronics markets. With more countries developing their design and production capabilities, the competition for world market share, let alone dominance, will be much keener that it has ever been.

The United States continues to lead the world in the development of new concepts, and there is no reason to think that will change. What will be challenged is our ability to cost-effectively produce products that are recognized worldwide for their quality. Then, and only then, will the United States be able to effectively compete in the worldwide marketplace.

Total Quality Emphasis

The total quality emphasis worldwide has focused attention where it should have always been—on the customer. Particularly in the government sector, little, if any, thought was given to what customers really wanted in terms of products and their capabilities. Through the advent of techniques such as quality function deployment (QFD), suppliers can now take the general needs criteria from the customer and formulate what needs to be done to design and build the product to satisfy the customer's needs. What we are really saying here is that we need to hear the voices of the customers as to their product needs.

To stop here with our discussion of total quality is to stop with only half of the voice of each customer heard. It is in the other half of the customer's voice that warranties have a profound impact. This other half of the customer's voice is known as customer satisfaction. Suppliers need to know what the customers want, but they also need to know what the customers got, or their perception of what they got.

In the military sector, customers must operate and use the items they purchase for as long as 20 years, and in some instances as in the case of the B-52 aircraft, much, much longer. The first few years, however, seem to be the most memorable to the customer. In the consumer sector the use duration is much shorter and the warranties are shorter. Current military warranties can be as long as 5 years, while typical consumer warranties do not exceed 1 year. In either case, the product's performance, quality, and reliability during this initial period of its life have a profound impact on the customer's overall satisfaction with the product.

Survival Strategy

To survive in the twenty-first century, suppliers must improve the reliability and performance of their products. As the Japanese have proven, customers are willing to pay more for a product if they perceive or have experienced an exceptional level of quality. Remembering that quality is defined to be giving the customer the best value within the constraints of their requirements, the value is much more than the price. Value includes the performance and reliability of the product.

Suppliers will be required to assume more risk to be successful in future years. The risk will have to be balanced internally by cost-cutting measures without degrading the quality of the products, and the availability and use of field data on their products, including customer satisfaction inputs. Field data fed back to the design community should result in certain quality problems being designed out early before the product is manufactured. Problems found during design are much cheaper to rectify then waiting until manufacture, and most especially waiting until the item is fielded. Intelligent collection and use of field data are essential to the overall improvement in customer satisfaction needed to maintain, or increase market share and be profitable.

Last, but certainly not least, to survive the future, a commitment of the total company to quality is necessary. It will no longer be good enough to have a few key managers, designers, process engineers, or field service engineers committed to quality improvement. The message must come down from top management that the company is committed to quality improvement, to wit, that the product will not be shipped unless it meets certain clear standards of quality. For example, Texas Instruments, Inc. of Dallas, Texas, winner of the Malcolm Baldrige National Quality Award in 1992, made very little quality progress until all TI employees were convinced that top management really believed that quality was most important. Until lines were shut down and products were reworked and even redesigned before they were shipped, at-large workers made no significant changes in how they did their jobs. It should also be noted, as Dr. Juran points out, that you must also be committed and able to hold the gains you have achieved in quality.

Let's relate these observations to warranties. A supplier who is able to effect quality improvement and remain price-competitive should also be able to take more risk and offer a better warranty than the competitors—better in terms of length, coverage, remedies, and even performance factors such as MTBF, response time, and availability and downtime. Guess what happens when you offer a better warranty with a competitive price? You guessed it—you sell more products, you have a lower warranty cost, and you have a good chance of surviving in the twenty-first century.

International Warranties

With the impending global marketplace come challenges for U.S. businesses relative to warranties. Currently, in many cases, warranties on consumer products purchased in the United States are not valid outside the United States. Similarly, warranties on consumer products purchased outside the United States may not be honored in the United States.

The question arises as to why this situation exists. If it is because of the lack of communication and cooperation, the challenge for U.S. companies becomes one of education, both to educate the international community and be educated by them. There is some precedence for this education, with good results. Apparently, within the European community, warranties for products purchased from manufacturers within any member country are valid in all member countries. Any U.S. businesses desiring to sell their products internationally would do well to investigate the methodology and mechanics of this European community agreement in an effort to develop a marketing strategy that comprehends the essence of this agreement.

If the reason why U.S. product warranties are not honored abroad is product quality, then U.S. suppliers need to embrace the concept of total quality discussed earlier. In order to survive in the global marketplace, quality will have to be the top priority of U.S.—and, in fact, any country's—companies.

With regard to foreign military sales (FMS), if a weapon system warranty is to be obtained for a FMS purchaser, the FMS purchaser's warranty requirements and the particular U.S. service's plan to obtain those requirements must be consistent and track the statute. If the FMS purchaser has requested unique warranty requirements, a more detailed plan may be required. Defense contractors must understand the mechanics of FMS contracting and communicate with the FMS customers to be sure that the warranty offered meets their needs. It is very important to determine at the outset if the FMS customer wants a weapon system warranty or a tailored warranty to meet specific needs.

Because there are now other sources for products that were once produced by the United States only, U.S. companies must research and understand how to effectively do business abroad. It may be that education alone can open the door to international markets. It may also be that confidence in the quality of U.S. products is the key. In either case, knowing that foreign countries do not necessarily do business as we do and that we are not a sole source is a must to compete internationally.

One final note. An area that requires extensive education on the part of U.S. companies is that of negotiation. We, by nature, are much more impatient regarding negotiation than virtually every other culture. In negotiating internationally, we must slow down and be patient and not assume that the negotiation can be completed in one day, or even a week. Patience is the byword for negotiating warranties.

Challenge

The author anticipates that 10-year or 150,000-mi warranties will be required to compete in the international automobile market within the next five years. In addition, required lifetime warranties on automobiles are a distinct possibility in the next 10 years for manufacturers to be competitive.

In the automobile business, and indeed in any business aspiring to be successful in the twenty-first century, new levels of quality and performance will be required to survive. Therein lies the challenge.

Bibliography

Air Force Regulation 70-11, *Weapon System Warranties,* December 1988.

Berke, T. M., and N. A. Zaino, "Warranties: What Are They? What Do They Really Cost?" *Annual Reliability and Maintainability Symposium Proceedings,* January 1991.

Brennan, J. R., "Life Cycle Cost," 3-day Short Course, Lear Jet, Wichita, KS, January 1993.

Brennan, J. R., and S. A. Burton, "Warranties: Planning, Analysis, and Implementation," 3-day Short Course, Virginia Polytechnic Institute and State University Northern Virginia Graduate Center, Falls Church, VA, June 1993.

Brennan, J. R., and J. T. Stracener, "Designing to Cost Effectiveness, Enhancing Quality," *Annual Reliability and Maintainability Symposium Proceedings,* January 1992.

Brennan, J. R., and S. A. Burton, "Warranties: Concept to Implementation," *Annual Reliability and Maintainability Symposium Proceedings,* January 1989.

Defense Federal Acquisition Regulation Supplement (DFARS) Subpart 246.770.

DOD Warranties, Effective Administration Systems Are Needed to Implement Warranties, U.S. General Accounting Office, report to the Secretary of Defense, September 1989.

Public Law 93-637, Magnuson-Moss Warranty-Federal Trade Commission Improvement Act, January 1975.

Reid, S., D. N. Isaacson, and J. R. Brennan, "Warranty Cost-Risk Analysis," *Annual Reliability and Maintainability Symposium Proceedings,* January 1991.

Reliability, Maintainability, and Supportability Guidebook, 2 ed., Society of Automotive Engineers, Inc., Warrendale, PA, June 1992.

United States Code, Section 2403 of Title 10.

The Warranty; A Useful Acquisition Tool, Publication 6506-4911, ARINC Research Corporation, March 1989.

Warranty Guidebook, Defense Systems Management College, Fort Belvoir, VA, October 1992.

Weapon System Warranty Conference, Society of Logistics Engineers, Fort Belvoir, VA, January 1993.

United States Government Documents

Warranty Guidebook (Oct. 1992), Defense Systems Management College, Fort Belvoir, VA 22060-5426.

Public Law 93-637, Library of Congress.

GAO/NSIAD Report No.89-57, Library of Congress or General Accounting Office.

Appendix

A

Magnuson-Moss Warranty—Federal Trade Commission Improvement Act

MAGNUSON-MOSS WARRANTY—FEDERAL TRADE COMMISSION IMPROVEMENT ACT

For Legislative History of Act, see p. 7702

PUBLIC LAW 93-637; 88 STAT. 2183

[S. 356]

An Act to provide minimum disclosure standards for written consumer product warranties; to define minimum Federal content standards for such warranties; to amend the Federal Trade Commission Act in order to improve its consumer protection activities; and for other purposes.

Be it enacted by the Senate and House of Representatives of the United States of America in Congress assembled, That:

This act may be cited as the "Magnuson-Moss Warranty—Federal Trade Commission Improvement Act".

TITLE I—CONSUMER PRODUCT WARRANTIES

DEFINITIONS

Sec. 101. For the purposes of this title:

(1) The term "consumer product" means any tangible personal property which is distributed in commerce and which is normally used for personal, family, or household purposes (including any such property intended to be attached to or installed in any real property without regard to whether it is so attached or installed).

(2) The term "Commission" means the Federal Trade Commission.

(3) The term "consumer" means a buyer (other than for purposes of resale) of any consumer product, any person to whom such product is transferred during the duration of an implied or written warranty (or service contract) applicable to the product, and any other person who is entitled by the terms of such warranty (or service contract) or under applicable State law to enforce against the warrantor (or service contractor) the obligations of the warranty (or service contract).

(4) The term "supplier" means any person engaged in the business of making a consumer product directly or indirectly available to consumers.

(5) The term "warrantor" means any supplier or other person who gives or offers to give a written warranty or who is or may be obligated under an implied warranty.

(6) The term "written warranty" means—

(A) any written affirmation of fact or written promise made in connection with the sale of a consumer product by a supplier to a buyer which relates to the nature of the material or workmanship and affirms or promises that such material or workmanship is defect free or will meet a specified level of performance over a specified period of time, or

(B) any undertaking in writing in connection with the sale by a supplier of a consumer product to refund, repair, replace, or take other remedial action with respect to such product in the event that such product fails to meet the specifications set forth in the undertaking,

which written affirmation, promise, or undertaking becomes part of the basis of the bargain between a supplier and a buyer for purposes other than resale of such product.

(7) The term "implied warranty" means an implied warranty arising under State law (as modified by sections 108 and 104 (a)) in connection with the sale by a supplier of a consumer product.

(8) The term "service contract" means a contract in writing to perform, over a fixed period of time or for a specified duration, services relating to the maintenance or repair (or both) of a consumer product.

(9) The term "reasonable and necessary maintenance" consists of those operations (A) which the consumer reasonably can be expected to perform or have performed and (B) which are necessary to keep any consumer product performing its intended function and operating at a reasonable level of performance.

(10) The term "remedy" means whichever of the following actions the warrantor elects:

(A) repair,

(B) replacement, or

(C) refund;

except that the warrantor may not elect refund unless (i) the warrantor is unable to provide replacement and repair is not commercially practicable or cannot be timely made, or (ii) the consumer is willing to accept such refund.

(11) The term "replacement" means furnishing a new consumer product which is identical or reasonably equivalent to the warranted consumer product.

(12) The term "refund" means refunding the actual purchase price (less reasonable depreciation based on actual use where permitted by rules of the Commission).

(13) The term "distributed in commerce" means sold in commerce, introduced or delivered for introduction into commerce, or held for sale or distribution after introduction into commerce.

(14) The term "commerce" means trade, traffic, commerce, or transportation—

(A) between a place in a State and any place outside thereof, or

(B) which affects trade, traffic, commerce, or transportation described in subparagraph (A).

(15) The term "State" means a State, the District of Columbia, the Commonwealth of Puerto Rico, the Virgin Islands, Guam, the Canal Zone, or American Samoa. The term "State law" includes a law of the United States applicable only to the District of Columbia or only to a territory or possession of the United States; and the term "Federal law" excludes any State law.

WARRANTY PROVISIONS

Sec. 102. (a) In order to improve the adequacy of information available to consumers, prevent deception, and improve competition in the marketing of consumer products, any warrantor warranting a consumer product to a consumer by means of a written warranty shall, to the extent required by rules of the Commission, fully and conspicuously disclose in simple and readily understood language the terms and conditions of such warranty. Such rules may require inclusion in the written warranty of any of the following items among others:

(1) The clear identification of the names and addresses of the warrantors.

(2) The identity of the party or parties to whom the warranty is extended.

(3) The products or parts covered.

(4) A statement of what the warrantor will do in the event of a defect, malfunction, or failure to conform with such written warranty—at whose expense—and for what period of time.

(5) A statement of what the consumer must do and expenses he must bear.

(6) Exceptions and exclusions from the terms of the warranty.

(7) The step-by-step procedure which the consumer should take in order to obtain performance of any obligation under the warranty, including the identification of any person or class of persons authorized to perform the obligations set forth in the warranty.

(8) Information respecting the availability of any informal dispute settlement procedure offered by the warrantor and a recital, where the warranty so provides, that the purchaser may be required to resort to such procedure before pursuing any legal remedies in the courts.

(9) A brief, general description of the legal remedies available to the consumer.

(10) The time at which the warrantor will perform any obligations under the warranty.

(11) The period of time within which, after notice of a defect, malfunction, or failure to conform with the warranty, the warrantor will perform any obligations under the warranty.

(12) The characteristics or properties of the products, or parts thereof, that are not covered by the warranty.

(13) The elements of the warranty in words or phrases which would not mislead a reasonable, average consumer as to the nature or scope of the warranty.

(b)(1)(A) The Commission shall prescribe rules requiring that the terms of any written warranty on a consumer product be made available to the consumer (or prospective consumer) prior to the sale of the product to him.

(B) The Commission may prescribe rules for determining the manner and form in which information with respect to any written warranty of a consumer product shall be clearly and conspicuously presented or displayed so as not to mislead the reasonable, average consumer, when such information is contained in advertising, labeling, point-of-sale material, or other representations in writing.

(2) Nothing in this title (other than paragraph (3) of this subsection) shall be deemed to authorize the Commission to prescribe the duration of written warranties given or to require that a consumer product or any of its components be warranted.

(3) The Commission may prescribe rules for extending the period of time a written warranty or service contract is in effect to correspond with any period of time in excess of a reasonable period (not less than 10 days) during which the consumer is deprived of the use of such consumer product by reason of failure of the product to conform with the written warranty or by reason of the failure of the warrantor (or service contractor) to carry out such warranty (or service contract) within the period specified in the warranty (or service contract).

(c) No warrantor of a consumer product may condition his written or implied warranty of such product on the consumer's using, in connection with such product, any article or service (other than article or service provided without charge under the terms of the warranty) which is identified by brand, trade, or corporate name; except that the prohibition of this subsection may be waived by the Commission if—

(1) the warrantor satisfies the Commission that the warranted product will function properly only if the article or service so identified is used in connection with the warranted product, and

(2) the Commission finds that such a waiver is in the public interest.

The Commission shall identify in the Federal Register, and permit public comment on, all applications for waiver of the prohibition of this subsection, and shall publish in the Federal Register its disposition of any such application, including the reasons therefor.

(d) The Commission may by rule devise detailed substantive warranty provisions which warrantors may incorporate by reference in their warranties.

(e) The provisions of this section apply only to warranties which pertain to consumer products actually costing the consumer more than $5.

DESIGNATION OF WARRANTIES

Sec. 103. (a) Any warrantor warranting a consumer product by means of a written warranty shall clearly and conspicuously designate such warranty in the following manner, unless exempted from doing so by the Commission pursuant to subsection (c) of this section:

(1) If the written warranty meets the Federal minimum standards for warranty set forth in section 104 of this Act, then it shall be conspicuously designated a "full (statement of duration) warranty".

(2) If the written warranty does not meet the Federal minimum standards for warranty set forth in section 104 of this Act, then it shall be conspicuously designated a "limited warranty".

(b) Sections 102, 103, and 104 shall not apply to statements or representations which are similar to expressions of general policy concerning customer satisfaction and which are not subject to any specific limitations.

(c) In addition to exercising the authority pertaining to disclosure granted in section 102 of this Act, the Commission may by rule determine when a written warranty does not have to be designated either "full (statement of duration)" or "limited" in accordance with this section.

(d) The provisions of subsections (a) and (c) of this section apply only to warranties which pertain to consumer products actually costing the consumer more than $10 and which are not designated "full (statement of duration) warranties".

FEDERAL MINIMUM STANDARDS FOR WARRANTY

Sec. 104. (a) In order for a warrantor warranting a consumer product by means of a written warranty to meet the Federal minimum standards for warranty—

(1) such warrantor must as a minimum remedy such consumer product within a reasonable time and without charge, in the case of a defect, malfunction, or failure to conform with such written warranty;

(2) notwithstanding section 108(b), such warrantor may not impose any limitation on the duration of any implied warranty on the product;

(3) such warrantor may not exclude or limit consequential damages for breach of any written or implied warranty on such product, unless such exclusion or limitation conspicuously appears on the face of the warranty; and

(4) if the product (or a component part thereof) contains a defect or malfunction after a reasonable number of attempts by the warrantor to remedy defects or malfunctions in such product, such warrantor must permit the consumer to elect either a refund for, or replacement without charge of, such product or part (as the case may be). The Commission may by rule specify

for purposes of this paragraph, what constitutes a reasonable number of attempts to remedy particular kinds of defects or malfunctions under different circumstances. If the warrantor replaces a component part of a consumer product, such replacement shall include installing the part in the product without charge.

(b)(1) In fulfilling the duties under subsection (a) respecting a written warranty, the warrantor shall not impose any duty other than notification upon any consumer as a condition of securing remedy of any consumer product which malfunctions, is defective, or does not conform to the written warranty, unless the warrantor has demonstrated in a rulemaking proceeding, or can demonstrate in an administrative or judicial enforcement proceeding (including private enforcement), or in an informal dispute settlement proceeding, that such a duty is reasonable.

(2) Notwithstanding paragraph (1), a warrantor may require, as a condition to replacement of, or refund for, any consumer product under subsection (a), that such consumer product shall be made available to the warrantor free and clear of liens and other encumbrances, except as otherwise provided by rule or order of the Commission in cases in which such a requirement would not be practicable.

(3) The Commission may, by rule define in detail the duties set forth in section 104(a) of this Act and the applicability of such duties to warrantors of different categories of consumer products with "full (statement of duration)" warranties.

(4) The duties under subsection (a) extend from the warrantor to each person who is a consumer with respect to the consumer product.

(c) The performance of the duties under subsection (a) of this section shall not be required of the warrantor if he can show that the defect, malfunction, of failure of any warranted consumer product to conform with a written warranty, was caused by damage (not resulting from defect or malfunction) while in the possession of the consumer, or unreasonable use (including failure to provide reasonable and necessary maintenance).

(d) For purposes of this section and of section 102(c), the term "without charge" means that the warrantor may not assess the consumer for any costs the warrantor or his representatives incur in connection with the required remedy of a warranted consumer product. An obligation under subsection (a)(1)(A) to remedy without charge does not necessarily require the warrantor to compensate the consumer for incidental expenses; however, if any incidental expenses are incurred because the remedy is not made within a reasonable time or because the warrantor imposed an unreasonable duty upon the consumer as a condition of securing remedy, then the consumer shall be entitled to recover reasonable incidental expenses which are so incurred in any action against the warrantor.

(e) If a supplier designates a warranty applicable to a consumer product as a "full (statement of duration)" warranty, then the warranty on such product shall, for purposes of any action under section 110(d) or under any State law, be deemed to incorporate at least the minimum requirements of this section and rules prescribed under this section.

FULL AND LIMITED WARRANTING OF A CONSUMER PRODUCT

Sec. 105. Nothing in this title shall prohibit the selling of a consumer product which has both full and limited warranties if such warranties are clearly and conspicuously differentiated.

SERVICE CONTRACTS

Sec. 106. (a) The Commission may prescribe by rule the manner and form in which the terms and conditions of service contracts shall be fully, clearly, and conspicuously disclosed.

(b) Nothing in this title shall be construed to prevent a supplier or warrantor from entering into a service contract with the consumer in addition to or in lieu of a written warranty if such contract fully, clearly, and conspicuously discloses its terms and conditions in simple and readily understood language.

DESIGNATION OF REPRESENTATIVES

Sec. 107. Nothing in this title shall be construed to prevent any warrantor from designating representatives to perform duties under the written or implied warranty: *Provided*, That such warrantor shall make reasonable arrangements for compensation of such designated representatives, but no such designation shall relieve the warrantor of his direct responsibilities to the consumer or make the representative a cowarrantor.

LIMITATION ON DISCLAIMER OF IMPLIED WARRANTIES

Sec. 108. (a) No supplier may disclaim or modify (except as provided in subsection (b)) any implied warranty to a consumer with respect to such consumer product if (1) such supplier makes any written warranty to the consumer with respect to such consumer product, or (2) at the time of sale, or within 90 days thereafter, such supplier enters into a service contract with the consumer which applies to such consumer product.

(b) For purposes of this title (other than section 104(a)(2)), implied warranties may be limited in duration to the duration of a written warranty of reasonable duration, if such limitation is conscionable and is set forth in clear and unmistakable language and prominently displayed on the face of the warranty.

(c) A disclaimer, modification, or limitation made in violation of this section shall be ineffective for purposes of this title and State law.

COMMISSION RULES

Sec. 109. (a) Any rule prescribed under this title shall be prescribed in accordance with section 553 of title 5, United States Code; except that the Commission shall give interested persons an opportunity for oral presentations of data, views, and arguments, in addition to written submissions. A transcript shall be kept of any oral presentation. Any such rule shall be subject to judicial review under section 18(e) of the Federal Trade Commission Act (as amended by section 202 of this Act) in the same manner as rules prescribed under section 18(a)(1)(B) of such Act, except that section 18(e)(3)(B) of such Act shall not apply.

(b) The Commission shall initiate within one year after the date of enactment of this Act a rulemaking proceeding dealing with warranties and warranty practices in connection with the sale of used motor vehicles; and, to the extent necessary to supplement the protections offered the consumer by this title, shall prescribe rules dealing with such warranties and practices. In prescribing rules under this subsection, the Commission may exercise any authority it may have under this title, or other law, and in addition it may require disclosure that a used motor vehicle is sold without any warranty and specify the form and content of such disclosure.

REMEDIES

Sec. 110. (a)(1) Congress hereby declares it to be its policy to encourage warrantors to establish procedures whereby consumer disputes are fairly and expeditiously settled through informal dispute settlement mechanisms.

(2) The Commission shall prescribe rules setting forth minimum requirements for any informal dispute settlement procedure which is incorporated into the terms of a written warranty to which any provision of this title applies. Such rules shall provide for participation in such procedure by independent or governmental entities.

(3) One or more warrantors may establish an informal dispute settlement procedure which meets the requirements of the Commission's rules under paragraph (2). If—

(A) a warrantor establishes such a procedure,

(B) such procedure, and its implementation, meets the requirements of such rules, and

(C) he incorporates in a written warranty a requirement that the consumer resort to such procedure before pursuing any legal remedy under this section respecting such warranty,

then (i) the consumer may not commence a civil action (other than a class action) under subsection (d) of this section unless he initially resorts to such procedure; and (ii) a class of consumers may not proceed in a class action under subsection (d) except to the extent the court determines necessary to establish the representative capacity of the named plaintiffs, unless the named plaintiffs (upon notifying the defendant that they are named plaintiffs in a class action with respect to a warranty obligation) initially resort to such procedure. In the case of such a class action which is brought in a district court of the United States, the representative capacity of the named plaintiffs shall be established in the application of rule 23 of the Federal Rules of Civil Procedure. In any civil action arising out of a warranty obligation and relating to a matter considered in such a procedure, any decision in such procedure shall be admissible in evidence.

(4) The Commission on its own initiative may, or upon written complaint filed by any interested person shall, review the bona fide operation of any dispute settlement procedure resort to which is stated in a written warranty to be a prerequisite to pursuing a legal remedy under this section. If the Commission finds that such procedure or its implementation fails to comply with the requirements of the rules under paragraph (2), the Commission may take appropriate remedial action under any authority it may have under this title or any other provision of law.

(5) Until rules under paragraph (2) take effect, this subsection shall not affect the validity of any informal dispute settlement procedure respecting consumer warranties, but in any action under subsection (d), the court may invalidate any such procedure if it finds that such procedure is unfair.

(b) It shall be a violation of section 5(a)(1) of the Federal Trade Commission Act (15 U.S.C. 45(a)(1)) for any person to fail to comply with any requirement imposed on such person by this title (or a rule thereunder) or to violate any prohibition contained in this title (or a rule thereunder).

(c)(1) The district courts of the United States shall have jurisdiction of any action brought by the Attorney General (in his capacity as such), or by the Commission by any of its attorneys designated by it for such purpose, to restrain (A) any warrantor from making a deceptive warranty with respect to a consumer product, or (B) any person from failing to comply with any requirement imposed on such person by or pursuant to this title or from violating any prohibition contained in this title. Upon proper showing that, weighing the equities and considering the Commission's or Attorney General's likelihood of ultimate success, such action would be in the public interest and after notice to the defendant, a temporary restraining order or preliminary injunction may be granted without bond. In the case of an action brought by the Commission, if a complaint under section 5 of the Federal Trade Commission Act is not filed within such period (not exceeding 10 days) as may be specified by the court after the issuance of the temporary restraining order or preliminary injunction, the order or injunction shall be dissolved by the court and be of no further force and effect. Any suit shall be brought in the district in which such person resides or transacts business. Whenever it appears to the court that the ends of justice require that other persons should be parties in the action, the court may cause them to be summoned whether or not they reside in the district in which the court is held, and to that end process may be served in any district.

(2) For the purposes of this subsection, the term "deceptive warranty" means (A) a written warranty which (i) contains an affirmation, promise, description, or representation which is either false or fraudulent, or which, in light of all of the circumstances, would mislead a reasonable individual exercising due care; or (ii) fails to contain information which is necessary in light of all of the circum-

stances, to make the warranty not misleading to a reasonable individual exercising due care; or (B) a written warranty created by the use of such terms as "guaranty" or "warranty", if the terms and conditions of such warranty so limit its scope and application as to deceive a reasonable individual.

(d)(1) Subject to subsections (a)(3) and (e), a consumer who is damaged by the failure of a supplier, warrantor, or service contractor to comply with any obligation under this title, or under a written warranty, implied warranty, or service contract, may bring suit for damages and other legal and equitable relief—

(A) in any court of competent jurisdiction in any State or the District of Columbia; or

(B) in an appropriate district court of the United States, subject to paragraph (3) of this subsection.

(2) If a consumer finally prevails in any action brought under paragraph (1) of this subsection, he may be allowed by the court to recover as part of the judgment a sum equal to the aggregate amount of cost and expenses (including attorneys' fees based on actual time expended) determined by the court to have been reasonably incurred by the plaintiff for or in connection with the commencement and prosecution of such action, unless the court in its discretion shall determine that such an award of attorneys' fees would be inappropriate.

(3) No claim shall be cognizable in a suit brought under paragraph (1)(B) of this subsection—

(A) if the amount in controversy of any individual claim is less than the sum or value of $25;

(B) if the amount in controversy is less than the sum or value of $50,000 (exclusive of interests and costs) computed on the basis of all claims to be determined in this suit; or

(C) if the action is brought as a class action, and the number of named plaintiffs is less than one hundred.

(e) No action (other than a class action or an action respecting a warranty to which subsection (a)(3) applies) may be brought under subsection (d) for failure to comply with any obligation under any written or implied warranty or service contract, and a class of consumers may not proceed in a class action under such subsection with respect to such a failure except to the extent the court determines necessary to establish the representative capacity of the named plaintiffs, unless the person obligated under the warranty or service contract is afforded a reasonable opportunity to cure such failure to comply. In the case of such a class action (other than a class action respecting a warranty to which subsection (a)(3) applies) brought under subsection (d) for breach of any written or implied warranty or service contract, such reasonable opportunity will be afforded by the named plaintiffs and they shall at that time notify the defendant that they are acting on behalf of the class. In the case of such a class action which is brought in a district court of the United States, the representative capacity of the named plaintiffs shall be established in the application of rule 23 of the Federal Rules of Civil Procedure.

(f) For purposes of this section, only the warrantor actually making a written affirmation of fact, promise, or undertaking shall be deemed to have created a written warranty, and any rights arising thereunder may be enforced under this section only against such warrantor and no other person.

EFFECT ON OTHER LAWS

Sec. 111. (a)(1) Nothing contained in this title shall be construed to repeal, invalidate, or supersede the Federal Trade Commission Act (15 U.S.C. 41 et seq.) or any statute defined therein as an Antitrust Act.

(2) Nothing in this title shall be construed to repeal, invalidate, or supersede the Federal Seed Act (7 U.S.C. 1551-1611) and nothing in this title shall apply to seed for planting.

(b)(1) Nothing in this title shall invalidate or restrict any right or remedy of any consumer under State law or any other Federal law.

(2) Nothing in this title (other than sections 108 and 104(a)(2) and (4)) shall (A) affect the liability of, or impose liability on, any person for personal injury, or (B) supersede any provision of State law regarding consequential damages for injury to the person or other injury.

(c)(1) Except as provided in subsection (b) and in paragraph (2) of this subsection, a State requirement—

(A) which relates to labeling or disclosure with respect to written warranties or performance thereunder;

(B) which is within the scope of an applicable requirement of sections 102, 103, and 104 (and rules implementing such sections), and

(C) which is not identical to a requirement of section 102, 103, or 104 (or a rule thereunder),

shall not be applicable to written warranties complying with such sections (or rules thereunder).

(2) If, upon application of an appropriate State agency, the Commission determines (pursuant to rules issued in accordance with section 109) that any requirement of such State covering any transaction to which this title applies (A) affords protection to consumers greater than the requirements of this title and (B) does not unduly burden interstate commerce, then such State requirement shall be applicable (notwithstanding the provisions of paragraph (1) of this subsection) to the extent specified in such determination for so long as the State administers and enforces effectively any such greater requirement.

(d) This title (other than section 102(c)) shall be inapplicable to any written warranty the making or content of which is otherwise governed by Federal law. If only a portion of a written warranty is so governed by Federal law, the remaining portion shall be subject to this title.

EFFECTIVE DATE

Sec. 112. (a) Except as provided in subsection (b) of this section, this title shall take effect 6 months after the date of its enactment but shall not apply to consumer products manufactured prior to such date.

(b) Section 102(a) shall take effect 6 months after the final publication of rules respecting such section; except that the Commission, for good cause shown, may postpone the applicability of such sections until one year after such final publication in order to permit any designated classes of suppliers to bring their written warranties into compliance with rules promulgated pursuant to this title.

(c) The Commission shall promulgate rules for initial implementation of this title as soon as possible after the date of enactment of this Act but in no event later than one year after such date.

TITLE II—FEDERAL TRADE COMMISSION IMPROVEMENTS

JURISDICTION OF COMMISSION

Sec. 201. (a) Section 5 of the Federal Trade Commission Act (15 U.S.C. 45)[1] is amended by striking out "in commerce" wherever it appears and inserting in lieu thereof "in or affecting commerce".

(b) Subsections (a) and (b) of section 6 of the Federal Trade Commission Act (15 U.S.C. 46(a), (b))[2] are each amended by striking out "in commerce" and inserting in lieu thereof "in or whose business affects commerce".

(c) Section 12 of the Federal Trade Commission Act (15 U.S.C. 52)[3] is amended by striking out "in commerce" wherever it appears and inserting in lieu thereof in subsection (a) "in or having an effect upon commerce," and in lieu thereof in subsection (b) "in or affecting commerce".

RULEMAKING

Sec. 202. (a) The Federal Trade Commission Act (15 U.S.C. 41 et seq.)[4] is amended by redesignating section 18 as section 21, and inserting after section 17 the following new section:

"Sec. 18. (a)(1) The Commission may prescribe—

"(A) interpretive rules and general statements of policy with respect to unfair or deceptive acts or practices in or affecting commerce (within the meaning of section 5(a)(1) of this Act), and

"(B) rules which define with specificity acts or practices which are unfair or deceptive acts or practices in or affecting commerce (within the meaning of such section 5(a)(1)). Rules under this subparagraph may include requirements prescribed for the purpose of preventing such acts or practices.

1. 15 U.S.C.A. § 45.
2. 15 U.S.C.A. § 46(a), (b).
3. 15 U.S.C.A. § 52.
4. 15 U.S.C.A. § 41 et seq.

Appendix B

Title 10, Sec. 2403, United States Code

§2403. Major weapon systems: contractor guarantees

(a) In this section:

(1) "Weapon system" means items that can be used directly by the armed forces to carry out combat missions and that cost more than $100,000 or for which the eventual total procurement cost is more than $10,000,000. Such term does not include commercial items sold in substantial quantities to the general public.

(2) "Prime Contractor" means a party that enters into an agreement directly with the United States to furnish part or all of a weapon system.

(3) "Design and manufacturing requirements" means structural and engineering plans and manufacturing particulars, including precise measurements, tolerances, materials, and finished product tests for the weapon system being produced.

(4) "Essential performance requirements," with respect to a weapon system, means the operating capabilities or maintenance and reliability characteristics of the system that are determined by the Secretary of Defense to be necessary for the system to fulfill the military requirement for which the system is designed.

(5) "Component" means any constituent element of a weapon system.

(6) "Mature full-scale production" means the manufacture of all units of a weapon system after the manufacture of the first one-tenth of the eventual total production or the initial production quantity of such system, whichever is less.

(7) "Initial production quantity" means the number of units of a weapon system contracted for in the first year of full-scale production.

(8) "Head of an agency" has the meaning given that term in section 2302 of this title [10 USCS § 2302].

(b) Except as otherwise provided in this section, the head of an agency may not after January 1, 1985, enter into a contract for the production of a weapon system unless each prime contractor for the system provides the United States with written guarantees that—

(1) the item provided under the contract will conform to the design and manufacturing requirements specifically delineated in the production contract (or in any amendment to that contract);

(2) the item provided under the contract, at the time it is delivered to the United States, will be free from all defects in materials and workmanship;

(3) the item provided under the contract will conform to the essential performance requirements of the item as specifically delineated in the production contract (or in any amendment to that contract); and

(4) if the item provided under the contract fails to meet the guarantee specified in clause (1), (2), or (3), the contractor will at the election of the Secretary of Defense or as otherwise provided in the contract—

(A) promptly take such corrective action as may be necessary to correct the failure at no additional cost to the United States; or

(B) pay costs reasonably incurred by the United States in taking such corrective action.

(c) The head of the agency concerned may not require guarantees under subsection (b) from a prime contractor for a weapon system, or for a component of a weapon system, that is furnished by the United States to the contractor.

(d) Subject to subsection (e)(1), the Secretary of Defense may waive part or all of subsection (b) in the case of a weapon system, or component of a weapon system, if the Secretary determines—

(1) that the waiver is necessary in the interest of national defense; or

(2) that a guarantee under that subsection would not be cost-effective.

The Secretary may not delegate authority under this subsection to any person who holds a position below the level of Assistant Secretary of Defense or Assistant Secretary of a military department.

(e)(1) Before making a waiver under subsection (d) with respect to a weapon system that is a major defense acquisition program for the purpose of section 139a of this title [10 USCS § 139a], the Secretary of Defense shall notify the Committees on Armed Services and on Appropriations of the Senate and House of Representatives in writing of his intention to waive any or all of the requirements of subsection (b) with respect to that system and shall include in the notice an explanation of the reasons for the waiver.

(2) Not later than February 1 of each year, the Secretary of Defense shall submit to the committees specified in paragraph (1) a report identifying each waiver made under subsection (d) during the preceding calendar year for a weapon system that is not a major defense acquisition program for the purpose of section 139a of this title [10 USCS § 139a] and shall include in the report an explanation of the reasons for the waivers.

(f) The requirement for a guarantee under subsection (b)(3) applies only in the case of a contract for a weapon system that is in mature full-scale production. However, nothing in this section prohibits the head of the agency concerned from negotiating a guarantee similar to the guarantee described in that subsection for a weapon system not yet in mature full-scale production. When a contract for a weapon system not yet in mature full-scale production is not to include the full guarantee described in subsection (b)(3), the Secretary shall comply with the notice requirements of subsection (e).

(g) Nothing in this section prohibits the head of the agency concerned from—

(1) negotiating the specific details of a guarantee, including reasonable exclusions, limitations and time duration, so long as the negotiated guarantee is consistent with the general requirements of this section;

(2) requiring that components of a weapon system furnished by the United States to a contractor be properly installed so as not to invalidate any warranty or guarantee provided by the manufacturer of such component to the United States;

(3) reducing the price of any contract for a weapon system or other defense equipment to take account of any payment due from a contractor pursuant to subclause (B) of subsection (b)(4);

(4) in the case of a dual source procurement, exempting from the requirements of subsection (b)(3) an amount of production by the second source contractor equivalent to the first one-tenth of the eventual total production by the second source contractor; and

(5) using written guarantees to a greater extent than required by this section, including guarantees that exceed those in clauses (1), (2), and (3) of subsection (b) and guarantees that provide more comprehensive remedies than the remedies specified under clause (4) of that subsection.

(h)(1) The Secretary of Defense shall prescribe such regulations as may be necessary to carry out this section.

(2) This section does not apply to the Coast Guard or to the National Aeronautics and Space Administration.

(Added Oct. 19, 1984, P. L. 98-525, Title XII, Part C, § 1234(a) in part, 98 Stat. 2601.)

Appendix

C

Defense Federal Acquisition Regulation Supplement Subpart 246.7

SUBPART 246.7—WARRANTIES

246.701 Definitions.

"Acceptance," as defined in FAR 46.701 and as used in this subpart and in the warranty clauses at FAR 52.246-17, Warranty of Supplies of a Noncomplex Nature; FAR 52.246.18, Warranty of Supplies of a Complex Nature; FAR 52.246-19, Warranty of Systems and Equipment Under Performance Specifications or Design Criteria; and FAR 52.246-20, Warranty of Services, includes the execution of an official document (e.g., DD Form 250, Material Inspection and Receiving Report) by an authorized representative of the Government.

"Defect," as used in this subpart, means any condition or characteristic in any supply or service furnished by the contractor under the contract that is not in compliance with the requirements of the contract.

246.702 General

(c) Departments and agencies shall establish procedures to track and accumulate data on warranty costs.

246.703 Criteria for use of warranties.

The use of warranties in the acquisition of weapon systems is mandatory (10 U.S.C. 2403) unless a waiver is authorized (see 246.770-8).

(b) *Cost.*
Contracting officers may include the cost of a warranty as part of an item's price or as a separate contract line item.

246.704 Authority for use of warranties.

The chief of the contracting office must approve use of a warranty, except in acquisitions for—

(1) Weapon systems (see 246.770);
(2) Commercial supplies or services (see FAR 46.709);
(3) Technical data, unless the warranty provides for extended liability (see 246.708);
(4) Supplies and services in fixed price type contracts containing quality assurance provisions that reference MIL-I-45208, Inspection System Requirement, or MIL-Q-9858, Quality Program Requirements; or
(5) Supplies and services in construction contracts when using the warranties that are contained in Federal, military, or construction guide specifications.

246.705 Limitations.

(a) Warranties in the clause at 252.246-7001, Warranty of Data, are also an exception to the prohibition on use of warranties in cost-reimbursement contracts.

246.706 Warranty terms and conditions.

(b)(5) *Markings.*
Use MIL Standard 129, Marking for Shipments and Storage, and MIL Standard 130, Identification Marking of U.S. Military Property, when marking warranty items.

246.708 Warranties of data.

Obtain warranties on technical data when practicable and cost effective. Consider the factors in FAR 46.703 in deciding whether to obtain warranties of technical data. Consider the following in deciding whether to use extended liability provisions—

(1) The likelihood that correction or replacement of the nonconforming data, or a price adjustment, will not give adequate protection to the Government; and
(2) The effectiveness of the additional remedy as a deterrent against furnishing nonconforming data.

246.710 Contract clauses.

(1) Use a clause substantially the same as the clause at 252.246-7001, Warranty of Data, in solicitations and contracts that include the clause at 252.227-7013, Rights in Technical Data and Computer Software, and there is a need for greater protection or period of liability than provided by other contract clauses, such as the clauses at—
    (i) FAR 52.246-3, Inspection of Supplies—Cost-Reimbursement;
    (ii) FAR 52.246-6, Inspection—Time-and-Material and Labor-Hour;
    (iii) FAR 52.246-8, Inspection of Research and Development—Cost-Reimbursement; and

- (iv) FAR 52.246-19, Warranty of Systems and Equipment Under Performance Specifications or Design Criteria.
- (2) Use the clause at 252.246-7001, Warranty of Data, with its Alternate I when extended liability is desired and a fixed price incentive contract is contemplated.
- (3) Use the clause at 252.246-7001, Warranty of Data, with its Alternate II when extended liability is desired and a firm fixed price contract is contemplated.

246.770 Warranties in weapon system acquisitions.

This section sets forth policies and procedures for use of warranties in contracts for weapon system production.

246.770-1 Definitions.

As used in this section—

- (a) "At no additional cost to the Government" means—
  - (1) At no increase in price for firm fixed price contracts;
  - (2) At no increase in target or ceiling price for fixed price incentive contracts (see also FAR 46.707); or
  - (3) At no increase in estimated cost or fee for cost-reimbursement contracts.
- (b) "Design and manufacturing requirements" means structural and engineering plans and manufacturing particulars, including precise measurements, tolerances, materials and finished product tests for the weapon system being produced.
- (c) "Essential performance requirements" means the operating capabilities and maintenance and reliability characteristics of a weapon system that the agency head determines to be necessary to fulfill the military requirement.
- (d) "Initial production quantity" means the number of units of a weapon system contracted for in the first program year of full-scale production.
- (e) "Mature full-scale production" means follow-on production of a weapon system after manufacture of the lesser of the initial production quantity or one-tenth of the eventual total production quantity.
- (f) "Weapon system" means a system or major subsystem used directly by the Armed Forces to carry out combat missions.
  - (1) The term includes, but is not limited to, the following (if intended for use in carrying out combat missions)—
    - (i) Tracked and wheeled combat vehicles;
    - (ii) Self-propelled, towed and fixed guns, howitzers and mortars;
    - (iii) Helicopters;
    - (iv) Naval vessels;
    - (v) Bomber, fighter, reconnaissance and electronic warfare aircraft;
    - (vi) Strategic and tactical missiles including launching systems;
    - (vii) Guided munitions;
    - (viii) Military surveillance, command, control, and communication systems;

(ix) Military cargo vehicles and aircraft;
(x) Mines;
(xi) Torpedoes;
(xii) Fire control systems;
(xiii) Propulsion systems;
(xiv) Electronic warfare systems; and
(xv) Safety and survival systems.

(2) The term does not include—
(i) Commercial items sold in substantial quantities to the general public (see FAR 15.804-3(c)); or
(ii) Spares, repairs, or replenishment parts; or
(iii) Related support equipment (e.g., ground-handling equipment, training devices and accessories, ammunition), unless an effective warranty would require inclusion of such items.

246.770-2 Policy.

(a) Under 10 U.S.C. 2403, departments and agencies may not contract for the production of a weapon system with a unit weapon system cost of more than $100,000 or an estimated total procurement cost in excess of $10 million unless—
(1) Each contractor for the weapon system provides the Government written warranties that—
(i) The weapon system conforms to the design and manufacturing requirements in the contract (or any modifications to that contract),
(ii) The weapon system is free from all defects in materials and workmanship at the time of acceptance or delivery as specified in the contract; and
(iii) The weapon system, if manufactured in mature full-scale production, conforms to the essential performance requirements of the contract (or any modification to that contract); and
(2) The contract terms provide that, in the event the weapon system fails to meet the terms of the above warranties, the contracting officer may—
(i) Require the contractor to promptly take necessary corrective action (e.g., repair, replace, and/or redesign) at no additional cost to the Government;
(ii) Require the contractor to pay costs reasonably incurred by the Government in taking necessary corrective action, or
(iii) Equitably reduce the contract price; or
(3) A waiver is granted under 246.770-8.

(b) Contracting officers may require warranties that provide greater coverage and remedies than specified in paragraph (a) of this subsection, such as including an essential performance requirement warranty in other than a mature full-scale production contract.

(c) When the contract includes an essential performance requirement warranty, the warranty must identify redesign as a remedy available to the Government.

(1) The period during which redesign must be available as a remedy shall not end before operational use, operational testing, or a combination of operational use and operational testing has demonstrated that the warranted item's design has satisfied the essential performance requirements.
(2) When essential performance requirements are warranted in contracts with alternate source contractors, do not include redesign as a remedy available to the Government under those contracts until the alternate source has manufactured the first ten percent of the eventual total production quantity anticipated to be acquired from that contractor (see 246.770-5).

246.770-3 Tailoring warranty terms and conditions.

(a) Since the objectives and circumstances vary considerably among weapon system acquisition programs, contracting officers must tailor the required warranties on a case-by-case basis. The purpose of tailoring is to get a cost-effective warranty in light of the technical risk, or other program uncertainties, while ensuring that the Government still acquires the basic warranties described in 246-770-2. Tailoring shall not be used as a substitute for acquiring a warranty waiver.
  (1) Tailoring may affect remedies, exclusions, limitations, and duration provided such are consistent with the specific requirements of this section (see also FAR 46.706).
  (2) Clearly relate the duration of any warranty to the contract requirements and allow sufficient time to demonstrate achievement of the requirements after acceptance.
  (3) Tailor the terms of the warranty, if appropriate, to exclude certain defects for specified supplies (exclusions) or to limit the contractor's liability under the terms of the warranty (limitations).
  (4) Structure broader and more comprehensive warranties when advantageous or narrow the scope when appropriate. For example, it may be inappropriate to require warranty of all essential performance requirements for a contractor that did not design the system.
(b) DoD policy is to exclude any terms that cover contractor liability for loss, damage, or injury to third parties from warranty clauses.
(c) Ensure acquisition of subsystems and components in a manner which does not affect the validity of the weapon system warranty.

246.770-4 Warranties on Government-furnished property.

Contracting officers shall not require contractors to provide the warranties specified in 246.770-2 on any property furnished the contractor by the Government, except for—

(a) Defects in installation;
(b) Installation or modification in such a manner that invalidates a warranty provided by the manufacturer of the property; or
(c) Modifications made to the property by the contractor.

246.770-5 Exemption for alternate source contractor(s).

Agency heads may exempt alternate source contractor(s) from the essential performance warranty requirements of 246.770-2(a)(1)(iii) until that contractor manufactures the first ten percent of its anticipated total production quantity.

246.770-6 Applicability to foreign military sales (FMS).

(a) The warranty requirements of 246.770-2 are not mandatory for FMS production contracts. DoD policy is to obtain the same warranties on conformance to design and manufacturing requirements and against defects in material and workmanship as it gets for U.S. supplies.
(b) DoD normally will not obtain essential performance warranties for FMS purchasers. However, where contracting officer cannot separately identify the cost for the warranty of essential performance requirements, the foreign purchaser shall be given the same warranty that the United States gets.
(c) If an FMS purchaser expressly requests a performance warranty in the letter of acceptance, the Government will exert its best efforts to obtain the same warranty obtained for U.S. equipment. Or, if specifically requested by the FMS purchaser, obtain a unique warranty.
(d) The costs for warranties for FMS purchasers may be different from the costs for such warranties for the Government due to factors such as overseas transportation and any tailoring to reflect the unique aspects of the FMS purchaser.
(e) Ensure that FMS purchasers bear all of the acquisition and administrative costs of any warranties.

246.770-7 Cost-benefit analysis.

(a) In assessing the cost effectiveness of a proposed warranty, perform an analysis which considers both the quantitative and qualitative costs and benefits of the warranty. Consider—
    (1) Costs of warranty acquisition, administration, enforcement, and user costs, and any costs resulting from limitations imposed by the warranty provisions;
    (2) Costs incurred during development specifically for the purpose of reducing production warranty risks;
    (3) Logistical and operational benefits as a result of the warranty as well as the impact of the additional contractor motivation provided by the warranty.
(b) Where possible, make a comparison with the costs of obtaining and enforcing similar warranties on similar systems.
(c) Document the analysis in the contract file. If the warranty is not cost effective, initiate a waiver request under 246.770-8.

246.770-8 Waiver and notification procedures.

(a) The Secretary of Defense has delegated waiver authority within the limits specified in 10 U.S.C. 2403. The waiving authority for the defense agencies is the Assistant Secretary of Defense (Production and Logistics). The waiving authority for the military departments is the Secretary of the department with authority to redelegate no lower than an Assistant Secretary. The waiving authority may waive one or more of the weapons system warranties required by 246.770-2 if—

(1) The waiver is in the interests of national defense; or

(2) The warranty would not be cost effective.

(b) Waiving authorities must make the following notifications or reports to the Senate and House Committees on Armed Services and Appropriations for all waivers—

(1) *Major Weapon Systems.* For a weapon system that is a major defense acquisition program for the purpose of 10 U.S.C. 2432, the waiving official must notify the committees in writing of an intention to waive one or more of the required warranties. Include an explanation of the reasons for the waiver in the notice. Ordinarily provide the notice 30 days before granting a waiver.

(2) *Other Weapon Systems.* For weapon systems that are not major defense acquisition programs for the purpose of 10 U.S.C. 2432, waiving officials must submit an annual report not later than February 1 of each year. List the waivers granted in the preceding calendar year in the report and include an explanation of the reasons for granting each waiver.

(3) *Weapon Systems Not in Mature Full-Scale Production.* Although a waiver is not required, if a production contract for a major weapon system not yet in mature full-scale production will not include a warranty on essential performance requirements, the waiving officials must comply with the notice requirements for major weapon systems.

(c) Departments and agencies shall issue procedures for processing waivers, notifications, and reports to Congress.

(1) Requests for waiver shall include—

(i) A brief description of the weapon system and its stage of production, e.g., the number of units delivered and anticipated to be delivered during the life of the program;

(ii) Identification of the specific warranty or warranties required by 246.770-2(a)(1) for which the waiver is requested;

(iii) The duration of the waiver if it is to go beyond the contract;

(iv) The rationale for the waiver (if the waiver request is based on cost-effectiveness, include the results of the cost-benefit analysis);

(v) A description of the warranties or other techniques used to ensure acceptable field performance of the weapon system, e.g., warranties, commercial or other guarantees obtained on individual components; and

(vi) Exercise date of the warranty option, if applicable.

(2) Notifications and reports shall include—

(i) A brief description of the weapon system and its stage of production; and

(ii) Rationale for not obtaining a warranty.

(3) Keep a written record of each waiver granted and notification and report made, together with supporting documentation such as a cost-benefit analysis, for use in answering inquiries.

Appendix

D

Army Regulation 700–139

Headquarters
Department of the Army
Washington, DC
10 March 1986

***Army Regulation 700–139**

Effective 10 April 1986

Logistics

Army Warranty Program Concept and Policies

This UPDATE printing publishes a new regulation which is effective 10 April 1986. It supersedes Army Warranty Program AR 702–13, dated 1 February 1981, in its entirety.

By Order of the Secretary of the Army
JOHN A. WICKHAM, JR.
General, United States Army
Chief of Staff

Official:

R. L. DILWORTH
Brigadier General, United States Army
The Adjutant General

Summary. This regulation on the policies, responsibilities, and procedures of the Army Warranty Program has been revised in consonance with 10 USC 2403 and the Federal Acquisition Regulation System. This regulation covers mandatory weapon system warranties and other nonmandatory warranties. This regulation assigns responsibilities, states acquisition policies, defines information requirements, covers fielding and execution procedures, and prescribes methods of compliance.

*This regulation supersedes AR 702–13, 1 February 1961.

Applicability. This regulation applies to the Active Army, the Army National Guard (ARNG), and the U.S. Army Reserve (USAR). This regulation applies to all Army acquired and managed items and non-Army acquired items used by the Army except—

a. Items purchased by nonappropriated funds.
b. Special intelligence property administered under AR 381–143.
c. Industrial production items.
d. Real property obtained or built by the Corps of Engineers.
e. Civil works activities of the Corps of Engineers.
f. Subsistence and clothing bought by the Defense Logistics Agency.
g. General Services Administration interagency motor pool vehicles and commercial design nontactical vehicles either purchased, leased, or rented.
h. Nonstandard equipment that is locally purchased.
i. Procurement of unprogrammed requirements in support of Special Operations Forces under AR 700–9.

Impact on New Manning System. This regulation does not contain information that affects the New Manning System.

Internal control systems. This regulation is subject to the requirements of AR 11–2. It contains internal control provisions and a checklist for conducting internal control reviews.

Supplementation. This regulation may be supplemented at the major Army command level, if required. One copy of each supplement will be furnished to HQDA (DALO-SMP-P), WASH DC 20310-0546.

Interim changes. Interim changes to this regulation are not official unless they are authenticated by The Adjutant General. Users will destroy interim changes on their expiration dates unless sooner superseded or rescinded.

Suggested improvements. The proponent agency of this regulation is the Office of the Deputy Chief of Staff for Logistics. Users are invited to send comments and suggested improvements on DA Form 2028 (Recommended Changes to Publications and Blank Forms) directly to HQDA (DALO-SMP-P), WASH DC 20310-0546.

Distribution. Distribution of this issue has been made in accordance with DA Form 12-9A requirements for 700 series publications. The number of copies distributed to a given subscriber is the number of copies requested in Block 506 of the subscriber's DA Form 12-9A. AR 700–139 distribution is B for Active Army, ARNG, and USAR.

Contents (listed by paragraph number)

Chapter 1
Introduction

Purpose • 1–1
References • 1–2
Explanation of abbreviations and terms • 1–3
Internal control • 1–4
Exemptions • 1–5

Chapter 2
Responsibilities

General • 2–1
Deputy Chief of Staff for Logistics (DCSLOG) • 2–2
Deputy Chief of Staff for Research, Development, and Acquisition (DCSRDA) • 2–3
Other Army Staff agency heads • 2–4
Executive agent • 2–5
Materiel developers • 2–6
Heads of gaining MACOMs • 2–7
Representatives of the Logistic Assistance Program • 2–8

Chapter 3
Statutory and Regulatory Requirements

General • 3–1
Weapon system warranties • 3–2
Nonweapon system warranties • 3–3
Warranties on locally procured items • 3–4
Warranties of technical data • 3–5
Program management documentation • 3–6

Chapter 4
Warranty Acquisition Policy and Procedures

Section I
Policy and Concepts
Policy • 4–1
Concepts • 4–2

Section II
Warranty Cost-Effectiveness and Assessment
Warranty cost-effectiveness • 4–3
Warranty assessments • 4–4

Section III
Reimbursements and Copayments
Army repair and reimbursement • 4–5
Copayments for prorata usage • 4–6

Section IV
Candidate Criteria and Warranty Coverage
Warranty candidates • 4–7
Warranty coverage • 4–8
Warranty duration • 4–9

Section V
Compatibility and Identification
Warranty compatibility with standard Army support systems • 4–10
Identification of warranty items • 4–11
Warranty technical bulletins • 4–12

Chapter 5
Warranty Information

General • 5–1
Central collection activity • 5–2
Warranty clause exchange • 5–3

Chapter 6
Warranty Fielding and Execution

Fielding of warranty items • 6–1
Warranty execution • 6–2

Chapter 7
Compliance

Materiel developer • 7–1
Gaining MACOMs • 7–2
Logistic Assistance Offices • 7–3
Executive agent • 7–4

Internal Control Review Checklists

1. Cost-Effectiveness Analysis and Payoff Assessment

Appendix A. References

Glossary

Chapter 1: Introduction

1.1. Purpose

This regulation prescribes Department of the Army (DA) policies and assigns responsibilities for the management and execution of the Army Warranty Program. This regulation governs warranties that apply to both centrally procured and locally procured items.

a. Centrally procured materiel warranties. This regulation establishes requirements and provides guidance for the management and performance of the Army Warranty Program for centrally procured materiel. The objectives of the Army Warranty Programs, as expressed within this regulation, are to—
  (1) Achieve and sustain a cost-effective warranty program for Army materiel.
  (2) Minimize user burden and promote user satisfaction.
  (3) Control warranty execution to assure maximum use and benefit from warranties.
  (4) Provide information for warranty administration, execution, and evaluation.
  (5) Achieve uniformity in managing and executing warranties.

b. Locally procured materiel warranties. Locally procured materiel warranties are governed by Federal Acquisition Regulation (FAR) 46.7.

1–2. References

Required and related publications are listed in appendix A.

1–3. Explanation of abbreviations and terms

Abbreviations and special terms used in this regulation are explained in the glossary.

1–4. Internal control

This registration contains an internal control review checklist for warranty cost effectiveness and payoff assessment. This checklist is located after the last chapter of this regulation.

1–5. Exemptions

The following programs are exempt from coverage in this regulation:

a. Reliability improvement warranties are defined in AR 702–3 and are more properly considered reliability improvement incentives.

b. Manufacturing dimensions and tolerance warranties when used for ammunition programs are exempt from coverage in this regulation. These warranties are, in effect, a delayed final inspection acceptance and are not executed outside of the manufacturing and load assembly environment.

c. Vehicle Safety Recall Campaign directives in compliance with section 1402, title 15, United States Code (15 USC 1402) implemented by AR 750–10 are exempt from this regulation.

Chapter 2 Responsibilities

2–1. General

Section 2403, title 10, United States Code (10 USC 2403) defines specific responsibilities of the Secretary of Defense for weapon system warranties. Army Weapon System waiver authority has been delegated to the Assistant Secretary of the Army (Research, Development, and Acquisition) by the Secretary of Defense. For the Army, the AR 10–series describes functions of the Army Staff and major Army commands (MACOMs). The AR 70–series and AR 700–series describe specific responsibilities in Army research and development, production engineering, product assurance, integrated logistics support, maintenance, and supply. Additional responsibilities for carrying out the Army Warranty Program are specified below.

2–2. Deputy Chief of Staff for Logistics (DCSLOG)

The DCSLOG has Army Staff responsibility for the management of the Army Warranty Program. The DCSLOG, in the Army Staff role, will—

a. Issue policy guidance for the technical requirements of warranties on both Army acquired items and non-Army acquired items used by the Army.
b. Issue policy guidance for the management of warranty compliance to the statutory requirements of 10 USC 2403, and regulatory requirements of FAR 46.7, Defense Federal Acquisition Regulation Supplement (DFARS) 46.7, and Army acquisition instructions.
c. Issue policy guidance pertaining to warranty management as part of integrated logistics support policy for the Army in AR 700–127 and in AR 700–129 where the Army is the lead service for Joint Service programs.
d. Issue policy guidance to institute data collection and reporting used to identify warranties, determine compliance, and facilitate warranty effectiveness evaluation.
e. Issue policy guidance to sustain compatibility between acquired warranties and the standard Army execution procedures. (See chap 6.)
f. Appoint an executive agent to carry out the DCSLOG responsibilities for the Army Warranty Program. (See para 2–5.)

2–3. Deputy Chief of Staff for Research, Development, and Acquisition (DCSRDA)

The DCSRDA will—

a. Issue policy guidance to assure appropriate warranty planning in materiel acquisition and management plans. (See 10 USC 2403, FAR 46.7, and DFARS 46.7.)
b. Issue policy guidance to assure that program acquisition strategy under AR 70–1 provides for warranty consideration within the acquisition plan and identifies funding for the acquisition, execution, and effectiveness evaluation of warranties.

2–4. Other Army Staff agency heads

Deputy Chief of Staff for Personnel (DCSPER), Deputy Chief of Staff for Operations and Plans (DCSOPS), Assistant Chief of Staff for Intelligence (ACSI),

Chief of Engineers (COE), and The Surgeon General (TSG) will implement the Army Warranty Program in their respective areas.

2–5. Executive agent

The Commanding General, U.S. Army Materiel Command (CG, AMC) will appoint the executive agent for the DCSLOG. The executive agent will—

a. Institute policy, determine compliance, and operate data collection and reporting methods in consonance with Headquarters, Department of the Army (HQDA) objectives.
b. Sustain compatibility of warranty execution methods with the standard Army supply and maintenance logistic support systems. (See para 4–10.)
c. Provide a weapon system warranty clause exchange service for materiel developers (MAT DEVs). (See para 5–3.)
d. Direct and control the central collection activity (CCA). (See para 5–2.)
e. Report annually to the DCSLOG on the Army Warranty Program and the effectiveness of the executive agency.

2–6. Materiel developers

The MAT DEV acquiring an item will determine applicability of 10 USC 2403 and FAR/DFARS regulatory requirements. The MAT DEV will—

a. Establish and maintain a command activity for managing warranted materiel.
b. Issue necessary supplemental policy and procedures that apply to the procurement of materiel warranties.
c. Identify the cost of the warranty.
d. Determine if the warranty is cost-effective (para 4–3) and apply the criteria of the 10 USC 2403, DFARS 46.7, and FAR 46.7 requirements as applicable. (See chap 3.)
e. Manage, monitor, and evaluate the effectiveness of procured warranties using approved supplements to this regulation (when required).
f. Perform annual, inprocess and postwarranty assessments to determine effectiveness and final payoff analyses of acquired warranties. (See para 4–4.)
g. Develop, in coordination with the U.S. Army Training and Doctrine Command (TRADOC), methods of executing warranties within the Integrated Logistic Support Plan (ILSP). Define duties of the gaining MACOMS for warranty execution in the Materiel Fielding Plan (MFP).
h. Sustain compatibility with the standard warranty execution procedures when gaining MACOMs perform the execution. (See chap 6.)
i. Coordinate with the gaining MACOM (or storage activity when applicable) for nonstandard execution procedures. (See para 6–2 b.)
j. Establish a warranty information data base for use by MAT DEV, gaining MACOM, Logistic Assistance Offices/logistic assistance representatives (LAO/LAR), and other Army activities. (See chap 5.)
k. Provide electronic mailbox (24-hour response) access to the central warranty information data base. (See para 5–2.)

l. Establish telephone access (24-hour HOTLINE) for resolution of execution problems or specific warranty questions from gaining MACOMs.

m. Provide warranty execution training as an integral part of materiel fielding/new equipment training with emphasis on geographic differences and unique organizational structures.

n. Assure warranty information, procedures, and other pertinent data is included in applicable technical bulletins/manuals and field technical documents.

o. Recoup from contracts (adjustments or reimbursed monies) for repair or replacements of warrantied items performed by Government activities, when Government repair or replacement is made in place of contractor repair.

2–7. Heads of gaining MACOMs

These officials will—

a. Assure that a warranty claim action (WCA) is filed for each failure of an item covered by a warranty.

b. Establish nonstandard execution procedures (para 6–2b) in coordination with MAT DEV when nonstandard procedures are acceptable to the gaining MACOM for their maintenance augmentation capability.

c. Provide suggestions or advice on the scope and methods of warranty execution as requested by the MAT DEV.

d. Recommend corrective action to the MAT DEV when published execution procedures prove unsatisfactory or result in extensive administrative burden.

e. Establish warranty acquisition, administration, and execution procedures for locally acquired items in compliance with FAR 46.7.

f. Include warranty functions within annual gaining MACOM budget submissions to provide for the administration and repair of warranted items.

g. Establish a warranty control office/officer (WARCO) at the MACOM level. MACOM WARCOs will—

(1) Review and coordinate with MAT DEV warranty execution procedures within MFPs, warranty technical bulletins (WTBs), and related warranty data to assure effective execution of warranties.

(2) Develop local written instructions for warranty execution and management within the MACOM.

(3) Direct the subordinate servicing WARCO function at the Directorate of Logistics (DOL) level for installation management organizations; at the intermediate—general support (INT–GS) for military organizations; at the State Maintenance Office level within the ARNG; and, at Army Reserve Commands for the USAR. Servicing WARCOs will—

(a) Execute warranties according to published procedures.

(b) Coordinate all warranty actions between its activities and commercial service sources (local dealer or manufacturer) and/or the MAT DEV as specified in WTBs. Such coordination does not include resolution of contractual issues.

(c) Maintain files and records as required to manage locally procured item warranties.

(4) Establish a coordinating subordinate WARCO function at MACOM determined levels such as corps, division, materiel management center, and area maintenance support activity when appropriate.

2–8. Representatives of the Logistic Assistance Program

LARs will provide advice and assistance to gaining MACOM WARCOs as part of its service interface under AR 700–4 between MAT DEV and gaining MACOMs. Representatives of the Logistic Assistance Program will—

a. Clarify warranty application/exclusions and warranty claim/report procedures upon WARCO or user request.
b. Assist WARCOs in developing local procedures for warranty administration.
c. Provide warranty information to users/WARCOs as a secondary source of information.
d. Provide specific assistance outlined in MFPs, technical and supply bulletins/manuals, and related documents for warranty management.

Chapter 3 Statutory and Regulatory Requirements

3–1. General

This chapter contains Army warranty policy concerning the United States code statute, FAR, and program documentation.

3–2. Weapon system warranties

a. Warranties will be acquired or waivers requested for items considered as weapon systems in accordance with 10 USC 2403 and the regulatory requirement of DFARS 46.7. Waiver authority is specified by DFARS 46.770–9.
b. Warranties for foreign military sales (FMS) are not required but may be elected by the FMS customer and may require special administration. (See DFARS 46.770–7.)

3–3. Nonweapon system warranties

a. Warranties will be acquired for items that are not covered by the 10 USC 2403 weapon system definition only when such warranties are cost-effective. These warranties will be acquired in accordance with the regulatory requirements of FAR 46.7 and DFARS 46.7 using the candidate selection criteria in this regulation. (See para 4–7.)
b. Commercial or trade practice warranties for centrally procured items will be acquired in accordance with FAR 46.7 when one of the following apply:
   (1) They are cost-effective and can be executed by the standard execution procedures.
   (2) They are cost-effective and can be executed by nonstandard execution procedures.

(3) The warranty cost cannot be severed by the MAT DEV from the item price to effect a price reduction for the item.

3–4. Warranties on locally procured items

Items that are locally procured will include warranties in accordance with FAR 46.7 only when they are cost-effective and executable by the item user. Administration and execution is the joint responsibility of the procuring activity and the item user. They must be jointly determined by local procedures prior to acquisition.

3–5. Warranties of technical data

DFARS 46.708 requires obtaining warranties for technical data whenever practicable and cost-effective. Computer software and computer software documentation are considered technical data under DFARS 52.227–7013, Rights in Technical Data.

3–6. Program management documentation

Program management documentation used for the Army System Acquisition Review Council (ASARC), inprocess review (IPR), or other decision authority reviews will include warranty consideration and plans as an integral of both the acquisition strategy and the integrated logistic support process of AR 1000–1.

Chapter 4 Warranty Acquisition Policy and Procedures

Section I
Policy and Concepts

4–1. Policy

The Army's policy for procuring warranties requires compliance with statutory and regulatory requirements. (See chap 3.) Cost-effectiveness and tailoring comprehensive coverage to fit the intended conditions and geographic locations of storage and use must be considered prior to contract award. Minimal tasks of execution to burden the operator, unit, or intermediate direct support maintenance organization is a major consideration in the tailoring of all warranties.

4–2. Concepts

Tailoring the warranty concept to fit the item and its intended use in a comprehensive manner with minimal impact on standard Army logistical procedures is the single most important aspect of the warranty acquisition process. Warranty tailoring is intended to protect the Army from the costs and frequency of systemic failures, enact responsive remedies for failures of significant operational impact, minimize or eliminate warranty execution tasks at the gaining MACOM, and become one of the methods used to require the contractor to fulfill the obligation of providing quality Army items. Two basic warranty concepts are frequently used; expected failures and failure-free.

a. Expected failure concept. The expected failure concept is based on the knowledge that the Army procures materiel to the minimum needs of the Army; therefore any design will include expected failures. A contract supplier should not be liable for those failures that are expected, but should be held liable for failures that exceed those that are expected. In order to use the expected failure concept, the Army and the supplier must have confidence in the reliability factors or specification data that yield a given quantity of failures that may occur during the warranty term.

(1) Items that utilize contractor depot or interim contract support for organic maintenance are readily adaptable to this concept since this occurs within the common contract.

(2) Items that are repaired at Army depots are also adaptable to this concept. However, the Army will incur additional cost for administration and the possibility of denied or disputed claims may increase.

(3) The use of this concept for INT–GS items requires the gaining MACOM to file a warranty claim for each failure. The MAT DEV will collect the claims. When the quantity threshold is reached, a contract remedy is then invoked for the excessive claims.

(4) The Army benefits from this warranty concept in several ways. The initial contract warranty is provided with little or no cost since the Army requires remedies only for excessive failures. When in operation, failure quantities which in sum are below the remedy threshold are an increase in product reliability and represent a cost avoidance. Likewise, total failure quantities in excess of the threshold are subject to the contract warranty remedy.

b. Failure-free concept. The failure-free concept requires a period of failure-free usage. Commercial and trade practice warranties are examples of this concept. Under this concept, each claim is subject to the contract remedy during the warranty term. Since failures may occur, the cost of the warranty will normally include the expense of repair or replacement that can be expected during the warranty term. This cost may be included in the item price and not identifiable as a separate cost.

(1) The Army's usage of failure-free warranties may occur when an item's reliability is unknown or unspecified, such as for a nondevelopment item. The use of the failure-free concept for items subordinate to the system level (para 4–7) may also be appropriate since they may not have individual indicators/recorders of usage such as an hour meter or odometer.

(2) Use of this concept must consider the cost of Army claim administration associated with the processing of each claim. This concept is often used in conjunction with the individual claim coverage (para 4–8a) of INT–GS and depot reparables/recoverables using the standard execution procedures. (See para 6–2a.)

(3) Use of this concept for items that have no INT–GS tasks (such as small arms weapons) is possible when used in conjunction with the systemic defect coverage (para 4–8b) as the method of contract remedy.

Section II
Warranty Cost-Effectiveness and Assessment

4–3. Warranty cost-effectiveness

MAT DEVs will institute procedures to determine the cost-effectiveness (AR 11–18 and AR 11–28) of warranties. Weapon system warranties require formal cost-effectiveness analysis. Nonweapon system warranty cost-effectiveness may be by either formal analysis or by documentation of rationale within the contract files.

a. Prior to negotiated procurement of an item warranty, a cost-effectiveness analysis is required to determine the value of the potential benefits received in comparison to the contract cost of the warranty plus the Army's cost of administration and execution.
b. Following receipt of formally advertised procurement bids, a cost-effectiveness analysis of the warranty is required to determine the value of the benefits in comparison to the contract cost (if separately priced) and the Army's cost of administration and execution.
c. Commercial or trade practice warranties for locally procured items require documentation of cost-effectiveness rationale within the contract files.
d. Commercial or trade practice warranties for centrally procured items require the cost-effectiveness analysis even when the warranty price is not severable from the item price.
   (1) This analysis is used to determine the value of the benefits (such as reduced maintenance or materiel cost) in comparison to the Army's cost of administration and execution plus any readiness-related cost. Additional float quantities purchased to effect a factory repair cycle time, response time cost, (in terms of equipment down-time), or other productive time lost attributable to the exercise of the warranty are readiness-related costs.
   (2) The cost-effectiveness analysis of a warranty, that is not severable from the item price, has a relation to storage, operation, and support costs and has three warranty execution possibilities. Execution of the warranty is implemented for the total remedies available, a selected group or level of remedies, or the execution of the warranty is not implemented because it is not cost-effective.
e. An internal control review checklist for cost-effectiveness determination is required for each contract warranty.

4–4. Warranty assessments

Assessments will be performed by MAT DEVs for warranties on an inprocess and final payoff basis.

a. Inprocess warranty assessments will be initiated concurrent with operation of the first item delivered under the contract. Subsequent inprocess assessments will be performed annually until all item warranties have expired and all claims settled and a final payoff assessment is compiled.
b. The assessments will, as a minimum, contain the identification of the con-

tract and contractor, a summary of claim activity during the period, and cumulative claim activity for the contract. Claim activity will include the claims submitted, honored, disputed, and denied and the value of each category. Denied claims will include reasons for denials such as false-pull (not deficient), abuse, or not covered by warranty. Denied and disputed claims will include a failure cause if applicable. In addition, an analysis will be performed to identify a proportional amount of the warranty cost to the value of warranty services/remedies received. A remarks section will include tasks or services that are considered desirable or undesirable based upon the claim frequency, failure mode, and value.

c. The final payoff assessment will evaluate the economic benefits derived from the warranty in comparison to the cost of corrective actions if there were no warranty. Cost avoidance as well as Government cost to administer the warranty must be considered. Nonmonetary benefits will be summarized and the inprocess assessments will be consolidated and summarized.

d. The warranty assessments will be used to determine warranty provisions and tasks for follow-on procurements for the item (and similar items) and the overall effectiveness of the item warranty.

e. An internal control review checklist for final payoff assessment determination is required for each contract warranty.

Section III
Reimbursements and Copayments

4–5. Army repair and reimbursement

Weapon system warranties will include a remedy that authorizes warranty repairs by the Army (or by Army contract) for which reimbursement will be made by the contractor. The reimbursement remedy is also required for nonweapon system warranties.

a. Contract recovery of expenses for materiel (parts), labor, and transportation incurred by the Army for repair or replacement of warranty items will be accomplished by contract refunds. Transportation expense recovery is necessary only when a warranty item's destination transportation cost exceeds the Army's normal repair facility destination cost for the item.

b. Contract recovery of gaining MACOM labor expenses (when part of the warranty coverage) will include labor expended for removal and replacement of items as well as the labor expended in the actual item repair. Labor rates used for contract computation will represent average Army maintenance labor costs for organic labor or the contractors burdened flat rate manual for labor. Maintenance allocation chart (MAC) labor hour standards will be used for computation. Summation of discrete labor hour tasks may be necessary to encompass the total repair effort.

c. Recovery of depot labor expenses will be limited to the labor expended in the item repair using the MAC or contractor labor hour standards. Labor rates used for contract computation will represent average Army depot labor rates.

d. Contract-recovered expenses will be refunded to a central DA fund for Operations and Maintenance, Army.

4–6. Copayment for prorata usage

a. Copayments for prorata usage are a payment of monies by the item owner, based on percentage of usage, to the item supplier (or representative) when a portion of warranty usage has occurred. Commercial tire and battery warranties are examples of prorata copayment warranties.

b. Copayments to contractors or dealers for prorata usage under an Army contract warranty will not be required from gaining MACOMs unless—

(1) The warranty items are covered by nonstandard warranty execution procedures negotiated as part of an MFP.

(2) The warranty items are commercial or trade practice items that are acceptable to the gaining MACOMs.

Section IV
Candidate Criteria and Warranty Coverage

4–7. Warranty candidates

Warranty candidates will be identified in accordance with the following criteria when the system or system subordinate items are the materiel to be procured:

a. *Weapon systems of 10 USC 2403.*

(1) Major systems identified in section 139*a*, title 10, United States Code (10 USC 139*a*).

(2) Systems not identified in 10 USC 139*a* but falling within the 10 USC 2403 definition.

(3) Items subordinate to the weapon system level that are—

(*a*) Within the cost criteria of DFARS 46.7.

(*b*) Depot reparable or depot recoverable by the maintenance and recovery codes of AR 700–82.

(*c*) Occur no lower than level 3 of the work breakdown structure (MIL-STD-881A; para 3.5.1) for prime mission hardware.

b. *Nonweapon systems.*

(1) Military and nonmilitary developed systems and system subordinate items that are listed (or proposed for listing) in Supply Bulletin 700–20, chapter 2.

(2) Are depot reparable or depot recoverable by the maintenance and recovery codes of AR 700–82.

(3) Occur no lower than the level directly below level 3 of the work breakdown structure (MIL-STD-881, para 3.5.1) for prime mission hardware.

4–8. Warranty coverage

Army warranties for centrally procured materiel will provide two coverages; individual item failure coverage and systemic defect coverage. Replacement assemblies may require both types of coverage. Commercial or trade practice warranties may be structured for both types of coverage. Pass-through warranties will be restricted in their usage.

a. Individual item failure coverage requiring individual warranty claim actions apply to MAC functions of maintenance or repair parts and special tool list (RPSTL) coded recovery functions that occur no lower than the INT–GS level for items and their subsidiary parts. Tasks for maintenance and recovery functions must be identified to the MAC or RPSTL for inclusion in the warranty but all of the identified functions may not be cost-effective for individual claim processing. The value of the function, as estimated by MAC labor hours, depot labor rates, and Army Master Data File (AMDF) part costs, must exceed the Army's cost of claim processing to be cost-effective as an individual warranty claim. When claim processing costs exceed the estimated value of the function, systemic defect coverage will be used instead of individual claim coverage.

b. Systemic defect coverage provides protection to the lowest level of impact or expense and requires a contract remedy that may cover all contract deliverables.

(1) When the contract warranty provisions include both individual item claims and systemic considerations, abnormal volume of WCAs against the particular part will initiate systemic contract remedies.

(2) When the contract warranty provisions do not include individual item claims, systemic failures will become evident by a significant number of product deficiency or other field reports. These include Quality Deficiency Reports, Equipment Improvement Recommendations, Report of Discrepancies, and other reports of product problems with the item.

(3) The MAT DEV, using the contract remedies, will arrange with the warranty contractor for an inventory-wide or total asset remedy when applicable. Replacements, recalls, or repairs will be coordinated with the gaining MACOM or depot as applicable. A comparable contract cost reduction may be appropriate in place of asset repair or replacement.

(4) An indepth analysis of the failure cause and a potential redesign may be necessary to prevent recurring failures.

(5) The term of coverage begins with the first contract item delivered and ends following the warranty expiration date of the last item delivered and includes all failures during the term.

c. System subordinate item contracts (para 4–7) for replacement assemblies or for assemblies integrated into systems as Government furnished equipment (GFE) may require both the systemic defect coverage and individual item failure coverage. This coverage is required for replacement items that received similar coverage under a system level warranty.

d. Commercial or trade practice warranties that extend coverage below the INT–GS level will be structured for individual item failure remedies for the INT–GS and depot level functions of maintenance and recovery and for systemic failure remedies at levels below INT–GS, when possible.

e. Pass-through warranties, which require the Army to seek remedies through vendors not directly under contract, will not be used on weapon system warranties. Commercial or trade practice warranties which have traditional subordinate pass-through warranties such as tires and batteries may be used.

4–9. Warranty duration

Warranty duration will be of sufficient time to provide a period for user operation that is proportional to the expected life of the item. The duration period is composed of two factors; average elapsed time prior to operation and operational use.

a. The average elapsed time factor is the period of time which occurs from the time of contract delivery (as evidenced by contract documented acceptance) until the item is placed into operation and includes all delays that may be normally expected prior to operational use. Included are transportation and storage delays, fielding in overseas geographic location delays, and delays planned when Government-furnished materiel is integrated into a higher weapon system.

b. The operational use factor is the period of time in actual operation that will prove the substantive quality of the item and the integrity of the manufacturing process. This period should be between 10 and 25 percent of the expected life and generally not less than 1 calendar year or 1 year of an equivalent usage rate in whatever units are best measured (for example, months, years, hours, miles, rounds).

c. When a warranty duration is computed for inclusion into a contract, the operational use factor is added to the average elapsed time factor to yield a single length duration which will be used for each delivered item.

(1) The duration period will start on the date of acceptance and each item will be identified with its unique expiration date.

(2) Items scheduled for long-term storage such as War Reserves or prepositioned stocks have the same duration as other items acquired for immediate operation. The average elapsed time factor will include the impact of long-term storage items and will result in either a longer duration period for all items under contract, or a comparable reduction in contract price for those items which have little likelihood of operational usage.

Section V
Compatibility and Identification

4–10. Warranty compatibility with standard Army support systems

Acquired warranties will sustain compatibility with the standard Army support systems. The item's support for the period under warranty will not differ from the follow-on support upon warranty expiration.

a. Storage and exercise of warranted items will not differ during the warranty from the item's postwarranty requirements.

b. Part support will operate within the Army's supply system for replacement parts. Urgent part support using direct shipment to Army maintenance facilities may be used for warranty items in the same manner that expedited shipment of nonwarranty items are used to fill urgent requisitions.

c. Warranty exhibits will be returned or disposed using the Army's disposal and retrograde return system. Specific items with return requirements or exhibit hold periods will be identified in the item's WTB and MFP.

d. WCAs will provide information to the MAT DEV and the warranty contractor in accordance with The Army Maintenance Management System (TAMMS), DA Pam 738–750 (nonaviation) and DA Pam 738–751 (aviation), TAMMS-A. Contract unique forms or information requirements will not be required when the gaining MACOM is expected to perform the standard Army execution procedures. (See para 6–2.)

e. Maintenance functions or work time figures of an item's MAC will not be changed to accommodate the warranty. The alignment of warranty coverage to maintenance levels and functions is to sustain normal support operations during the warranty period with the support that will follow warranty expiration. During the course of normal support operations, it may become necessary to move, subdivide, or combine MAC functions to accommodate the Army's support needs. The MAT DEV will attempt to realign the warranty with the MAC changes if cost-effectiveness and execution can be sustained. If contract changes cannot be accomplished, some functions may be unilaterally excluded from execution for not complying with the changed MAC.

f. Warranty remedies should not be any less responsive than normal Army maintenance methods to sustain readiness. Contract warranty provisions will be defined for responsiveness in terms of time (hours, days, weeks) between notification and resolution of a warranty claim.

4–11. Identification of warranty items

The Army's standard execution procedures (para 6–2) are based on a free flow return of failed items to the claiming level of maintenance. The passiveness of the procedures require obvious markings to allow for identification screening at the claiming levels. Therefore, warranty identification/data plates and package marking is a contract requirement and will be added to Army documentation as a requirement. In some instances, an item may be excluded from individual claim coverage and may be included under systemic claim coverage because physical size, shape, or material makes identification markings impossible. In other instances, logbook or historical record data may be used for identification purposes for items of a system level warranty.

a. Warranty information/data plates, as specified in contracts and Army documentation, will be applied to the system hardware and to depot and INT–GS reparable/recoverables that comprise the system covered by the warranty. The data plate marking requirement will comply with MIL-STD-130 and the following requirements:

(1) Minimum information markings will include "WARRANTY ITEM," "WTB XXXXX" (unique number), and "EXPIRES XX/XX" (unique date/rate). The "EXPIRES XX/XX" will be expressed as numeric month slash numeric year or usage rate (for example, hours, miles). Characters will be either white or black to obtain maximum contrast to the background. Bar coding of the warranty data and the national stock number (NSN), contract number, and contractor Federal Supply Code of Manufacturers (FSCM) number is desired but not mandatory.

(2) Background marking requirements will provide alternating blue and neutral (natural color of material) 45 degree diagonal stripes of equal

width. The width of each stripe will be approximately equal to the character height. Blue color will approximate FED-STD-595, color number 35250.

b. Warranty package/packaging markings will comply with MIL-STD-129 for size and information marked. In addition, background markings as specified above for data plates will be applied to packages/packaging.

c. Expiration date/usage marked on plates and packages will be applied at contract acceptance for each item and will be that period defined as the warranty duration period (para 4–9) or usage rate equivalent.

d. Shipping and release documents will identify warranty items in the appropriate form area or remarks section to inform the receiver of the existence of warranty materiel. This applies to items being issued for use and items being evacuated for repair.

e. Computer programs that appear on a visual display will include a notice of warranty coverage on one of the introductory screens of the program. The warranty coverage details will be presented within the program.

4–12. Warranty technical bulletins

Warranty provisions for execution will be published in a WTB prepared in accordance with MIL-M-63034 (TM) in sufficient time to provide draft copies for MFP coordination and final copies concurrent with materiel fielding. WTBs may, by necessity, be a contract deliverable item in order to be available for MFP coordination. When WTBs are contract prepared, they will be procured by a contract line item number (CLIN) and exhibit.

Chapter 5 Warranty Information

5–1. General

Warranty information will be collected and shared by MAT DEVs and gaining MACOM organizations to document and improve warranties and their benefits using a CCA as the combined data base.

a. MAT DEVs will collect and provide to the CCA, information on each warranty for centrally procured items. This information will include—

(1) NSN, nomenclature, and model numbers.
(2) Contract number, contractor name, and FSCM.
(3) Warranty publication (for example, WTB) number and date.
(4) Serial, lot, or registration number range (when applicable).
(5) Warranty duration (time in months)
(6) Warranty usage limits (hrs/miles/km).
(7) Start date of first item warranty period.
(8) End date of last item warranty expiration.
(9) Contract cost of warranty (sum and per unit) and contract item cost.
(10) Subordinate (pass-through) warranties if applicable.
(11) Special warranty provisions or conditions.

b. MAT DEVs will collect, collate, and automate WCAs submitted from all sources and provide information access to the CCA and annual reports to gaining MACOMs.

(1) Data or information expected to be gathered from gaining MACOMs or activities will be limited to WCA data of DA Pam 738–750 and DA Pam 738–751.

(2) Data or information gathered as part of nonstandard execution procedures will, as a minimum, provide the same data elements gathered by TAMMS/TAMMS-A WCAs. Special data collection programs such as sampling data by AR 750–37 and interim contractor support program (ICS) data are examples of special information sources.

(3) Contract status reports (DI-A-1025) provide an alternate or corroborate means of acquiring and verifying claim data.

(4) Data or information gathered within systems integration programs or depot operations as warranty claims will, as a minimum, provide the same data elements gathered by TAMMS/TAMMS-A WCAs.

5–2. Central collection activity

The executive agent directed CCA serves as a central source of automated warranty information. The CCA serves to—

a. Collect information gathered by the MAT DEV and operate a combined data base.

b. Publish listings/reports for warranty information users (MAT DEVs, WARCOs, LAOs).

(1) WARCO addresses and an index of warranty items published in DA Pam 738–750 and DA Pam 738–751.

(2) Warranty Highlighter (information letter) published periodically.

(3) Annual summary reports of MACOM and WARCO activity for annual compliance analysis.

c. Provide access to the data base as an electronic mailbox for queries of individual warranty coverage specifics within 24 hours from receipt of request.

5–3. Warranty clause exchange

A weapon system warranty clause exchange service will be provided by the legal office of the executive agent for MAT DEV. This service will supply copies of existing warranty clauses upon request and does not supplant legal or procurement review requirements of the MAT DEV. The purpose of the service is to proliferate successful clauses used for procurement of weapon system warranties under 10 USC 2403.

Chapter 6
Warranty Fielding and Execution

6–1. Fielding of warranty items

Warranty items will be fielded in accordance with appropriate materiel release, fielding or transfer documents noting specific warranty requirements in the MFP.

a. Survey of local service sources.

(1) Concurrent with MFP negotiation, the materiel fielder will conduct a survey of capacity and capability of local service sources where utilization of these sources is planned.

(2) Concurrent with fielding, the gaining MACOM WARCO will resurvey the service sources to confirm servicing capability and capacity.

b. WTBs will be provided with the MFP and each item when required. In addition, WTBs will be distributed by pinpoint publication distribution methods.

c. The materiel fielding team (MFT) will provide WTB copies to the gaining MACOM and coordinating/servicing WARCOs, to the MACOM LAO, and to local LAO/LAR.

d. WTB details of coverage and execution will be explained by the MFT to WARCOs, LAOs, and LARs.

e. Gaining MACOM budget programming to accomplish maintenance, supply, and retrograde recovery tasks associated with warranty execution must encompass the potential of Army repair and contract recovery of expenses.

6–2. Warranty execution

a. Standard Army execution procedures (SAEPs). SAEPs fulfill the requirements of minimum burden, compatibility with the normal Army logistical support system, and uniformity/simplicity of administration. The basic premise of these procedures is to support the item during the warranty in the same manner as that which occurs in postwarranty ownership.

(1) TAMMS/TAMMS-A procedures contain instructions and forms for completing WCAs. Contract unique forms or procedures are not used for WCAs.

(2) Individual WCAs do not occur below the INT–GS level.

(3) Supply support and retrograde recovery flow through the normal Army logistical systems.

(4) Storage and exercise requirements for warranty items do not differ from the Army's postwarranty requirements.

b. Nonstandard execution procedures. Nonstandard execution procedures are not used when execution is to be performed by gaining MACOMs except when—

(1) The MACOM agrees to perform nonstandard execution for maintenance augmentation as part of the MFP.

(2) The methods of collection in AR 750–37 are utilized and no unique burden is applied to the gaining MACOM.

(3) Interim contractor support agreements provide for the WCAs and execution.

(4) Warranties do not extend beyond the wholesale level and are executed by the MAT DEV or depot system.

(5) Warranties are included as part of a local procurement.

c. Warranty exhibits. Warranty exhibits (as specified in the WTB) utilize the standard retrograde return system when execution is performed by gaining MACOMs.

(1) Preservation and safeguarding of warranty exhibits are a priority task of the gaining MACOM to protect the contract remedies of the Army.
(2) Evacuation of warranty exhibits conform to the MFP and WTB instructions. Storage of exhibits is provided by the gaining MACOM pending disposition instructions from the MAT DEV.
(3) Disposition instructions are furnished (by the MAT DEV) to the gaining MACOM within 30 calendar days of the MAT DEV notification of WCA receipt.

Chapter 7
Compliance

7–1. Materiel developer

MAT DEV compliance will be accomplished by—

a. Inspector general review of compliance to the statutory requirements.
b. Executive agent review of the annual inprocess and postwarranty assessments, command visits such as the command logistics review team (CLRT) reports, and compliance visits.
c. Internal control provisions of warranty checklists.

7–2. Gaining MACOMs

Gaining MACOM compliance will be accomplished by—

a. Inspector general review of compliance to this regulation and MACOM supplementation when applicable.
b. Executive agent review of claim summaries, command visits such as the CLRT reports, and compliance visits.

7–3. Logistic Assistance Offices

LAO/LAR compliance will be accomplished by executive agent (command LAO) annual review of data repository and procedures review for each LAO support office.

7–4. Executive agent

Executive agent compliance will be accomplished by the DCSLOG, using the annual reports of the executive agency.

Internal Control Review Checklist

Task:	Army Warranty Program	*Organization:*
Subtask:	Warranty Cost and Benefits:	*Action officer:*
This checklist:	Cost-Effectiveness Analysis and Payoff Assessment	*Reviewer:*
Event cycle 1:	Warranty Cost-Effectiveness (C-E)	*Date completed:*

Step #1: Submit warranty C-E analysis summary to MACOM HQ for approval.

Risk: Contract warranty will be procured without appropriate MACOM HQ approval of C-E analysis.

Control objective: Assure that each MACOM review the warranty C-E analysis and that the analysis receives MACOM approval.

Control technique: Establish written procedures for coordination of all warranty C-E analyses.

TEST QUESTION	*Response* Yes	No	NA	Remarks[1]
Are all warranty cost-effectiveness analyses submitted through proper channels to MACOM HQ for approval?				

Step #2: C-E analyses are conducted in conformance with approved policies and procedures.

Risk: C-E analyses will not be conducted with approved policies and procedures.

Control objective: Assure that policies and procedures are established for the conduct of C-E analyses.

Control technique: Establish current, written policies and procedures for conducting C-E analyses.

TEST QUESTION	*Response* Yes	No	NA	Remarks[1]
Have current policies and procedures for conducting C-E analyses been written and disseminated?				

[1]Provide reference to documentation or explanation for response.

Step #3: Disseminate the most current, approved C-E analysis model as a source and reference document.

Risk: MACOM approved C-E analysis model is not readily available for use by subordinate contracting activities as a method of C-E analysis preparation.

Control objective: Assure that the C-E analysis model is published and used.

Control technique: Publish and update the C-E analysis model as the method for analysis of warranty C-E.

TEST QUESTION	*Response*			
	Yes	No	NA	Remarks[1]
Has the C-E analysis model been published, updated, and disseminated to contract activities of the MACOM?				

Step #1: Document contract file with C-E analysis and rationale for warranty decision.

Risk: Contract warranties will be procured without appropriate documentation of the contract files.

Control objective: Assure that each contact be documented with C-E analysis and rationale for warranty decision.

Control technique: Establish written procedures for inclusion of C-E analysis and rationale for warranty decision within contract files.

TEST QUESTION	*Response*			
	Yes	No	NA	Remarks[1]
Have contract files been documented with the C-E analysis and warranty decision?				

Step #2: Document contract file with payoff assessment of each warranty prior to contract close-out.

Risk: Contracts will be closed out without an assessment of the final warranty benefits.

Control objective: Assure that each contract be documented with an assessment of the final warranty benefits.

Control technique: Establish written procedures for inclusion of a warranty payoff assessment for each contract warranty prior to contract close-out.

TEST QUESTION	*Response*			
	Yes	No	NA	Remarks[1]
Have the contract files been documented with warranty payoff assessment prior to contract close-out?				

The above-listed internal controls provide reasonable assurance that Army resources are adequately safeguarded. I am satisfied that if the above listed controls are fully operational, the internal controls for this subtask throughout the Army are adequate.

Signed by: James B. Emahiser
Functional Proponent

I have reviewed this subtask within my organization and have supplemented the prescribed internal control review checklist when warranted by unique environmental circumstances. The controls prescribed in this checklist, as amended, are in place and operational for my organization (except for the weaknesses described in the attached plan, which includes schedules for correcting the weaknesses).

Operating manager (signature)

This checklist must be used within 120 days of initial publication and every 2 years thereafter. See AR 11–2 for specific requirements of the Internal Control Program.

Appendix A
References

Section I
Required Publications

DA Pam 738–750
The Army Maintenance Management System (TAMMS). (Cited in paras 4–10*d*, 5–1*b*, 5–2*b*, and 6–2*a*.)

DA Pam 738–751
The Army Maintenance Management System—Aviation (TAMMS-A). (Cited in paras 4–10*d*, 5–1*b*, 5–2*b*, and 6–2*a*.)

DFARS 46.7
Defense Federal Acquisition Regulation Supplement, Warranties. (Cited in paras 2–2*b*, 2–3*a*, 2–6*d*, 3–2*a*, 3–3*a*, and 4–7*a*.)

DFARS 46.708
Defense Federal Acquisition Regulation Supplement, Warranties of Technical Data. (Cited in para 3–5.)

DFARS 46.770–7
Defense Federal Acquisition Regulation Supplement, Applicability of FMS. (Cited in para 3–2*b*.)

DFARS 46.770–9
Defense Federal Acquisition Regulation Supplement, Waiver and Notification Procedures. (Cited in para 3–2*a*.)

DFARS 52.227–7013
Defense Federal Acquisition Regulation Supplement, Rights in Technical Data. (Cited in para 3–5.)

DI-A-1025
Data Item Description for Contract Status Reports. (Cited in para 5–1*b*.)

FAR 46.7
Federal Acquisition Regulation, part 46.7. (Cited in paras 1–1*b*, 2–2*b*, 2–3*a*, 2–6*d*, 2–7*e*, 3–3*a*, 3–3*b*, and 3–4.)

FED-STD 595
Federal Standard 595, Colors. (Cited in para 4–11*a*(2).)

MIL-STD-129J
Marking for Shipment and Storage. (Cited in para 4–11*b*.)

MIL-STD-130F
Identification Marking of US Military Property. (Cited in para 4–11*a*.)

MIL-STD-881A
Work Breakdown Structures for Defense Materiel Items. (Cited in paras 4–7*a* and 4–7*b*.)

Supply Bulletin 700–20
Army Adopted/Other Items Selected for Authorization/List of Reportable Items. (Cited in para 4–7*b*.)

Section II
Related Publications

A related publication is merely a source of additional information. The user does not have to read it to understand this regulation.

AR 11–2
Internal Control Systems

AR 11–18
The Cost Analysis Program

AR 11–28
Economic Analysis and Program Evaluation for Resource Management

AR 70–1
System Acquisition Policy and Procedures

AR 381–143
Logistic Policies and Procedures

AR 700–4
Logistic Assistance Program

AR 700–9
Policies of the Army Logistics System

AR 700–82
Joint Regulation Governing the Use and Application of Uniform Source, Maintenance and Recoverability Codes

AR 700–127
Integrated Logistic Support

AR 700–129
Joint Integrated Logistic Support

AR 702–3
Reliability Improvement Warranties

AR 750–10
Modification of Materiel and Issuing Safety-of-Use Messages and Commercial Vehicle Safety Recall Campaign Directives

AR 750–37
Sample Data Collection

AR 1000–1
Basic Policies for System Acquisition

MIL-M-63034(TM)
Manual Technical: Warranty Technical Bulletins, Preparation of

Glossary

Section I
Abbreviations

AMC
U.S. Army Materiel Command

AMDF
Army Master Data File

ASARC
Army System Acquisition Review Council

ASCI
Assistant Chief of Staff for Intelligence

ARNG
Army National Guard

CCA
central collection activity

CLIN
contract line item number

CLRT
command logistics review team

COE
Chief of Engineers

DA
Department of the Army

DCSLOG
Deputy Chief of Staff for Logistics

DCSOPS
Deputy Chief of Staff for Operations and Plans

DCSPER
Deputy Chief of Staff for Personnel

DCSRDA
Deputy Chief of Staff for Research, Development, and Acquisition

DFARS
Defense Federal Acquisition Regulation Supplement

DOL
Director of Logistics

FAR
Federal Acquisition Regulation

FMS
foreign military sales

FSCM
Federal Supply Code of Manufacturers

GFE
Government-furnished equipment

HQDA
Headquarters, Department of the Army

ILSP
Integrated Logistic Support Plan

INT–GS
intermediate—general support

IPR
Inprocess review

LAO
Logistic Assistance Office

LAR
logistic assistance representative

MAC
maintenance allocation chart

MACOM
major Army command

MAT DEV
materiel developer

MFP
materiel fielding plan

MFT
materiel fielding team

NSN
national stock number

RPSTL
repair parts and special tool list

SAEP
standard Army execution procedure

TAMMS
The Army Maintenance Management System

TAMMS-A
The Army Maintenance Management System—Aviation

TRADOC
U.S. Army Training and Doctrine Command

TSG
The Surgeon General

USAR
U.S. Army Reserve

WARCO
warranty control office/officer

WCA
warranty claim action

WTB
warranty technical bulletin

Section II
Terms

Centrally procured
Procurements made in support of materiel managed by the national inventory control point.

Cost-effective
A warranty that has tangible and intangible benefits which exceed the cost to procure, administer, and execute the warranty.

Cost-effectiveness analysis
An analysis between cost to procure, administer, and execute a warranty compared to the value of tangible and intangible benefits received.

Executable
The ability of the Army to put into operation a contract warranty and make warranty claims within the normal functions of maintenance and supply operations.

Execution
The process of carrying out the Army's right to apply for contract remedies under a warranty, such as making warranty claims.

Exhibit
A part or group of parts that are the residual materiel remaining from a warranty repair action. Broken or failed assemblies or the parts of assemblies that have failed may qualify as exhibits based on the WTB specifies.

Federal Supply Code of Manufacturers
A five-position code assigned to organizations that manufacture or maintain design control for items purchased, used, and cataloged by agencies of the Federal Government.

Gaining MACOM
The field command that receives materiel and puts the materiel into operational use.

Item
Item used in this regulation indicates procured materiel.

Materiel developer
Command or agency responsible for research, development, and production of a system in response to approved requirements.

Systemic failure
A classification for failure which occur with a frequency, pattern, or sameness to indicate a logical regularity of occurrence.

Warranty
Warranty, as used in this regulation (and FAR 46.7), means a promise or affirmation given by a contractor to the Government regarding the nature, usefulness, or condition of the supplies or services furnished under the contract.

Appendix

Secretary of the Navy Instruction 4330.17

DEPARTMENT OF THE NAVY
OFFICE OF THE SECRETARY
WASHINGTON, D.C. 20350-1000

SECNAVINST 4330.17
SO-4 (CBM)
18 SEP 1987

SECNAV INSTRUCTION 4330.17

From: Secretary of the Navy
Subj: NAVY POLICY ON USE OF WARRANTIES
Ref: (a) Navy Acquisition Regulations Supplement (NARSUP) SUBPART 46.72
(b) Federal Acquisition Regulation (FAR) SUBPART 46.7
(c) DoD FAR Supplement (DFARS) SUBPART 46.7

1. *Purpose.* To ensure that the Department of Navy (DON) obtains and administers warranties that enhance the quality, reliability and performance of systems, subsystems and materials.
2. *Scope.* This instruction applies to all Fleet, Fleet Marine Force and Shore activities involved in logistics support for DON systems, subsystems and materials.
3. *Policy.* It is DON Policy to:
   a. Ensure that Navy obtains warranties for:
      (1) all weapons systems used directly by the armed forces. This applies to weapons systems which will have a unit cost greater than $100,000, or for which the eventual total procurement cost will be more than $10,000,000, unless such warranties are determined not to be cost effective.
      (2) all other supplies and services (i.e., non-weapons systems), when the contracting officer determines that obtaining a warranty is ad-

vantageous to the Government. Such warranties must equal or exceed the requirements of DFARS 46.770.

b. Ensure that Systems are established for:
   (1) reporting failed items under warranty
   (2) user return of warranted products
   (3) collecting and analyzing actual warranty use and claim data

4. *Action.* Addressees will implement and provide copies of implementing instructions to ASN (Shipbuilding and Logistics) Contract Business Management within 120 days. Detailed directives should address the issues presented in reference (a).

   a. The Chief of Naval Operations will:
      (1) establish procedures to ensure that warranties are obtained for:
         (*a*) weapons systems meeting the thresholds specified here.
         (*b*) all other supplies and services (i.e., non-weapons systems) per references (b) and (c).
      (2) establish procedures to ensure maximum use of warranted products before expiration of the warranty periods.
      (3) establish a customer/user notification system which provides for feedback information on failed items under warranty, minimizing reporting requirements of fleet activities and maintenance personnel.
      (4) develop procedures for immediate issuance of credit to the end item user, when appropriate, when requisitioned products under warranty are found to be defective upon installation.
      (5) develop a system for collecting actual warranty use and claim data, and for performing an analysis of the data on an annual basis with the first analysis to be performed on 30 June following implementation of this instruction, and annually each June thereafter. Provide copies of annual warranty data analyses to the Assistant Secretary of the Navy (Shipbuilding & Logistics) (ASN(S&L)) within 60 days of the end of each annual analysis period.

   b. The Commandant of the Marine Corps will develop warranty policy for Marine Corps acquisitions, and establish procedures for processing warranty claims.

   c. The Comptroller of the Navy will ensure that procedures are available to collect funds under warranties and that those funds are properly credited to the appropriate accounts.

EVERETT PYATT
ASSISTANT SECRETARY OF THE NAVY
(SHIPBUILDING AND LOGISTICS)

Distribution:

SNDL	A2A	(NAVCOMPT, OGC)
	A3	(Chief of Naval Operations)
	A6	(Headquarters, U. S. Marine Corps)

Copy to:
SNDL A1 (Assistant Secretary of the Navy (Shipbuilding and Logistics))
(Assistant Secretary of the Navy (Financial Management))
SNDL FL1 (Naval Data Automation Command) (Code-813, only) (20 copies)

Stocked:
CO, NAVPUBFORMCEN
5801 Tabor Avenue
Philadelphia, PA 19120-5099 (100 copies)

Appendix

F

Air Force Regulation 70-11

DEPARTMENT OF THE AIR FORCE
Headquarters US Air Force
Washington DC 20330-5000

AF REGULATION 70-11

1 December 1988

Acquisition Management

Weapon System Warranties

Weapon system warranties (WSW) provide the Air Force ways to motivate contractors to design, produce, and deliver quality weapon systems as well as a means to correct defects for which the contractor is responsible. This regulation provides policy and procedures, and assigns responsibilities for acquiring, administering, and reporting of WSWs. It is to be used in conjunction with the Federal Acquisition Regulation (FAR), Subpart 46.7, Warranties, and the Department of Defense FAR Supplement (DFARS) and the Air Force FAR Supplement (AFFARS) thereto.

This regulation applies to all Air Force activities engaged in the acquisition and administration of WSWs. This includes subsidiaries or affiliated agencies for which the US Air Force has support responsibilities, such as US Air Force Reserve (USAFR) and Air National Guard (ANG) units and members.

No. of Printed Pages:20
OPR:SAF/AQCS(Lt Col J. Avon)
Approved by: Brig Gen John D. Slinkard
Writer-Editor: R. M. Downey
Distribution: F

Section A—General Information

1. Terms Explained:
   - *a. Action Point.* The organization or individual responsible for all actions necessary to investigate a problem under the Service/Deficiency Reporting System and to determine possible courses of action to resolve it.
   - *b. Cost-Benefit Analysis.* An analytical procedure used to determine if a warranty is cost effective by analyzing both the qualitative and quantitative costs and benefits of the warranty.
   - *c. Defect.* As used in this regulation, a defect is any condition or characteristic in supplies or services furnished under a contract that does not conform to the contract provisions. (Also see Department of Defense Federal Acquisition Regulation Supplement (DFARS) Subpart 46.701).
   - *d. Design and Manufacturing Requirements.* Structural and engineering plans and manufacturing particulars, including precise measurements, tolerances, materials, and finished product tests for the weapon system being produced (Also see DFARS, Subpart 46.770-1).
   - *e. Essential Performance Requirements.* Measurable, verifiable, trackable, and enforceable operating capabilities and/or maintenance and reliability characteristics of a weapon system that are determined to be necessary for the system to fulfill the military requirement for which the system is designed (Also see DFARS, Subpart 46.770-1).
   - *f. Foreign Military Sales.* That portion of United States security assistance authorized by the Foreign Assistance Act of 1961, as amended, and the Arms Export Control Act of 1976, as amended. This assistance differs from the Military Assistance Program and the International Military Education and Training Program in that the recipient provides reimbursement for defense articles and services transferred.
   - *g. Implementing Command.* The Air Force command responsible for developing and acquiring the weapon system, subsystem, or item of equipment.
   - *h. Initial Production Quantity.* The number of units of a weapon system contracted for in the first program year of full-scale production (Also see DFARS, Subpart 46.770-1). Full-scale production means that production beyond low-rate-initial-production.
   - *i. Mature Full-Scale Production.* As used in this regulation, production of a weapon system after manufacture of the lesser of the initial production quantity or one-tenth of the eventual total production quantity (Also see DFARS, Subpart 46.770-1).
   - *j. Product Performance Agreement.* A form of warranty, guarantee, or incentive used in a government contract to achieve or improve product performance or supportability in the operational environment.
   - *k. Program Manager.* The single Air Force manager (system program director, program or project manager, system program manager or item manager) during any specific phase of the acquisition life cycle.
   - *l. Supporting Command.* The command assigned responsibility for providing logistics support for weapon systems, subsystems, and equipment; it assumes program management and engineering responsibility from the implementing command.

m. *Using Command.* The command assigned responsibility for operating, employing, and deploying Air Force weapon systems, subsystems, and equipment in the conduct of training or actual combat operations.

n. *Warranty.* A promise or affirmation given by the contractor to the government regarding the nature, usefulness, or condition of the supplies or performance of services furnished under the contract.

o. *Weapon System.* As used in this regulation and consistent with DFARS, Subpart 46.770-1, a system or major subsystem used directly by the Armed Forces to carry out combat missions. By way of illustration, the term "weapon system" includes, but is not limited to the following, if intended for use in carrying out combat missions: tracked and wheeled combat vehicles; self-propelled, towed, and fixed guns, howitzers and mortars; helicopters; naval vessels; bomber, fighter, reconnaissance and electronic warfare aircraft; strategic and tactical missiles including launching systems; guided munitions; military surveillance, command, control, and communication systems; military cargo vehicles and aircraft; mines; torpedoes; fire control systems; propulsion systems; electronic warfare systems; and safety and survival systems. This term does not include related support equipment, such as ground-handling equipment, training devices and accessories thereto; or ammunition, unless an effective warranty for the weapon system would require inclusion of such items. This term does not include commercial items sold in substantial quantities to the general public as described at FAR Subpart 15.804-3(c).

p. *Weapon System Warranties Manager.* The office (or individual), designated by the program manager, responsible for management and administration of a specific contractual warranty.

q. *Weapon System Warranties Plan.* A plan containing program warranty strategy, terms of the warranty, and administration and enforcement requirements.

2. WSW Program Objectives. The objectives of the WSW Program are to:
   a. Develop and acquire warranties that:
      (1) Motivate the contractor to ensure product quality and performance.
      (2) Continue contractor responsibility and involvement beyond the delivery date and for the entire warranty period.
      (3) Are easy to manage and administer, such that there is no disruption to existing military systems and procedures.
      (4) Are enforceable.
      (5) Are affordable in relation to potential benefits.
   b. Provide standard procedures for identifying, reporting, tracking, and correcting defects and failures covered by a contractual warranty, including performance measurement and tracking of weapon systems, equipment, and items.
   c. Minimize the need for new and costly warranty data tracking systems and related manpower resources to administer contract warranties.
3. Background Information. The US Air Force has long recognized the importance of ensuring product quality in fielded weapon systems and equipment through the use of warranties, guarantees, and various performance

incentive arrangements, i.e., product performance agreements (PPA). The Defense Procurement Reform Act of 1985 (Title 10, United States Code, Section 2403) reemphasized the importance of warranties by enacting permanent statutory requirements for warranting weapon systems that are entering mature full-scale production. This regulation provides the basic policies, procedures, and responsibilities to effectively implement WSW requirements.

Section B—Basic Policies and Procedures

4. Applicability and Scope:
   - *a. General Information.* The focus of this regulation is on weapon system warranties. As such, these warranties must be acquired, administered, and reported as required by this regulation and DFARS, Subpart 46.7 as supplemented, and meet the objectives stated in paragraph 2. When determined to be in the best interest of the government, the policies and procedures set forth in this regulation may be used when acquiring nonstandard FAR or commercial warranties for other items and services.
   - *b. Weapon Systems.* All weapon systems entering into mature full-scale production with a unit weapon system cost of more than $100,000, or for which the eventual total procurement cost is in excess of $10,000,000 must be covered by a weapon system warranty in accordance with 10 U.S.C. 2403, as implemented by DFARS, Subpart 46.7 and the Air Force supplement thereto, unless a waiver is approved (paragraph 15). The prime contractor must guarantee that the weapon system provided under the production contract will:
     - (1) Conform to the design and manufacturing requirements specifically delineated in the production contract (or any amendment to that contract).
     - (2) Be free from all defects in materials and workmanship at the time it is delivered to the United States Government.
     - (3) Conform to the essential performance requirements of the item as specifically delineated in the production contract (or any amendment to that contract).

     The three guarantees described above combine to form a WSW as required by 10 U.S.C. 2403. The first two guarantees (subparagraphs (1) and (2)) warrant the contract specification while the third guarantee (subparagraph (3)) warrants selected performance parameters. Also, program offices may require warranties that provide greater coverage and remedies than specified in this regulation, such as including EPR warranty coverage in other than a mature full-scale production contract.
   - *c. Other Items and Services.* All other items and services not meeting the weapon system definition or the cost thresholds identified in b above, may be covered by a warranty as specified in FAR Subpart 46.7 and supplements thereto. The policies and procedures in this regulation may be used.

d. Technical Data. In accordance with DFARS, Subpart 46.708 and the Air Force Supplement thereto, warranty of technical data should be obtained whenever practical and cost effective.

e. Foreign Military Sales (FMS). DFARS, Subpart 46.770-7, Applicability to FMS, provides DOD policy on acquiring WSWs for all weapon systems procured for FMS requirements.

f. Foreign Military Acquisitions. AFFARS, Subpart 46.770-92, Foreign Military Acquisitions, provides Air Force guidance on acquiring WSWs for all weapon systems procured through a foreign government for United States Government use. The same guidance must be followed on acquiring WSWs from foreign sources except when the acquisition laws of the host country apply (e.g., country-to-country memorandum of agreement may require the laws of the host country from which the weapon system is being procured to apply).

5. WSW Planning:

a. The intent to use warranties must be established early in the acquisition cycle. Acquisition plans supporting Demonstration and Validation (DEM/VAL) and Full Scale Development (FSD) efforts should address the applicability of and planning for obtaining a WSW on production contracts. If feasible, a WSW should be considered for use during FSD. A sample warranty provision that places a contractor on notice that a WSW will be required on the production contract must be placed in DEM/VAL Request for Proposal (RFP). The provision at this time may be only a framework that identifies the essential performance requirements that will be warranted and the remedies to be invoked for the correction of defects. A more complete model provision that sets forth all the warranty terms and conditions must be included in the FSD RFP. Results from the DEM/VAL and FSD phases should be used to tailor warranty requirements for the production phase. When the government requires contractors to propose upon government-developed clauses at the time of FSD or later, they may also propose alternatives that the government will evaluate. Warranty strategy should be reassessed periodically. Attachment 1 shows how warranty-related activities interface with the system life cycle.

b. A determination to apply a warranty to a weapon system impacts not only the implementing command but also the supporting and using commands and responsible contract administration office. Therefore, the program office must prepare and coordinate a WSW plan with supporting and using commands, responsible contract administration office, and the program manager must approve it as required in Attachment 2 to this regulation. The WSW plan may be an attachment to other program office-generated plans. The program manager must approve the WSW plan within 6 months after award of the FSD contract, and the program manager must update it for the initial and follow-on production contracts.

c. A warranty plan is also required when a warranty is to be contractually acquired for nonweapon systems, items, or services that will require using, supporting, and participating command support to administer and enforce the warranty. The government also requires a warranty plan for FMS and foreign military acquisitions when a WSW will be acquired.

6. Cost-Benefit Analysis (CBA). It is DOD policy to obtain only cost-effective WSWs. Therefore, a CBA must be done to determine whether the contemplated WSW, which will be in the production contract, is cost effective. A CBA must be done, even though the contractor may propose a "no-cost" WSW, to compare the government's cost of administering and enforcing the WSW to the potential benefits to be derived from the proposed WSW. DFARS, Subpart 46.770-8 contains DOD policy concerning WSW CBA and AFR 173-15, paragraph 4-7, provides Air Force guidance for conducting the CBA, as well as when the Air Force should accomplish and update the CBA. The Product Performance Agreement Center (PPAC) has also developed a computer model to help program offices in doing the WSW CBA. When accomplishing the CBA, the information contained in the Warranty Activity Report, if available, must be considered (paragraph 16d).
7. Pricing Considerations. In addition to the guidelines contained in the Armed Service Pricing Manual, the following guidelines should be followed when developing the WSW price:
   a. The price of a warranty may include reasonable costs, but not profit, for the repair or replacement of a minimum number of random or predicted failures, caused by manufacturing or material and workmanship defects, in the early production lots. The contractor shall be totally responsible for the repair or replacement of any failures in excess of the minimum accepted level. Costs, which may be recognized include: engineering and manufacturing labor, parts and materials, shipping and handling, etc.

   b. The price of a warranty shall not include any costs for redesign efforts. The contractor, in most cases, is paid to design an item, and once the contractor signs up to the design and performance specification, should bear all costs in meeting those specifications. It would not be unreasonable, however, to allow for redesign costs, if the design was specified or developed by the government or another contractor.

   c. The price of a warranty may include reasonable warranty administration and warranty data costs. Reasonable costs may include the salary of a warranty manager, information management systems to collect and report warranty data, engineering costs related to the evaluation of warranty data, etc. Maintenance agreements for repair or replacement are not warranties and should be priced separately.

   d. When pricing a warranty, the rule-of-thumb approach should be avoided. Instead, a bottom-up approach, or if adequate data is available, cost-estimating relationships, or a combination of the two approaches should be used to price a warranty. Contractors shall be required to provide detailed breakdowns of their warranty price, and all proposed costs must be fully justified. Engineering assistance should be obtained in evaluating the proposed costs, especially in analyzing the predicted number of random failures, and the estimated costs to repair such failures.

   e. In deriving a cost-effective warranty, the Contracting Officer may provide for certain exclusions and limitations in the terms of the WSW clause which must be considered when pricing the warranty (DFARS, Subpart 46.770-3).

8. Essential Performance Requirements (EPR):

   a. The ability to affect design to achieve EPRs decreases rapidly as a weapon system moves through the research and development phases. Therefore, contractors must be alerted early in the acquisition cycle, ideally no later than the DEM/VAL phase, that the government intends to require a performance warranty under the production contract. This may be accomplished by identifying EPRs or goals in the DEM/VAL contract. It is expected that these goals or requirements will continue to be refined as the weapon system proceeds through development. EPRs must be consistent with the operational effectiveness and suitability requirements as well as pertinent performance and support parameters and goals. These requirements will be specified in statements of need (SON), depot support requirements documents (DSRD), and system operational requirements documents (SORD). For the weapon system production contract, the EPRs subject to warranty must be described in the contract specifications. EPRs must be identified in the WSW plan required by paragraph 5b.

   b. An EPR should be selected based on operational performance requirements for which compliance cannot be determined with certainty prior to or during acceptance testing. Such requirements include reliability, maintainability, and availability. The contractor's compliance with these requirements may only be determined as a result of field operations in the environment in which the weapon system is required to operate. Select those EPRs which will be measured during the normal field operation and maintenance of the weapon system, as defined in the operation and maintenance concepts, using existing field performance data collection systems.

   c. Generally, system level EPRs should be selected for warranty coverage rather than EPRs that apply to lower tiers such as component or line replaceable unit (LRU). For example, if a prime contractor is providing the total weapon system, the EPRs selected for warranty coverage should be at the weapon system level rather than at the subsystem or lower level. On the other hand, the government often provides major subsystems as government furnished equipment (GFE) to the prime contractor. In this instance, the EPRs may be at the major subsystem level versus the LRU level or component level.

   The GFE prime contractor would be responsible for the warranty on the GFE item.

   d. In accordance with AFFARS, Subpart 46.770-4, authority for designating EPRs is delegated to commanders of major commands with power to redelegate. EPRs must be coordinated with the using and supporting commands prior to their incorporation into any contract.

9. Waivers and Deviations to Specification Requirements:

   a. Prior to approval of any proposed waiver or deviation to a particular requirement set forth in the contract specification, a written evaluation of the impact of the proposed waiver or deviation on the WSW EPRs must be accomplished. In no event will a waiver or deviation be approved that releases the contractor from responsibility for complying with the WSW EPR unless a Secretarial waiver is approved in accordance with Section C. In such cases, an equitable adjustment to the contract price and other terms and conditions of the contract must be accomplished.

b. To ensure the government's approval of a waiver, deviation, or engineering change proposal request submitted until MIL-STD-480A, the WSW must require an impact statement to be submitted with the request by the contractor.

10. Remedies:

a. Each WSW must clearly describe the remedies available to the government to correct a manufacturing defect or performance failure covered under the WSW. For example, remedies for EPR breaches should provide for the immediate restoral of combat capability (through use of consignment spares), no cost ECPs (to fix the breach), and subsequent retrofit of new designs at no cost to the government. As a minimum, the WSW must provide for the remedies specified in DFARS, Subpart 46.770-2(a)(2), which are described below:

(1) Require the contractor to promptly take such corrective action as necessary (e.g., repair, replace or redesign) at no cost to the United States Government.

(2) Require the contractor to pay costs reasonably incurred by the United States in taking necessary corrective action (i.e., government repair).

(3) Equitably reduce the contract price (e.g., may be appropriate when combat capability is not affected).

b. When contractor repair or replacement is stipulated as an authorized remedy, also stipulate the required turn around time from contractor receipt of the defective or failed item to contractor shipment or government acceptance of the repaired or replacement serviceable item. Also stipulate the government remedy should the contractor fail to meet the guaranteed turnaround time, e.g., consignment spares.

c. If government repair is authorized, clearly identify the conditions, limitations and exclusions that may apply. Also indicate how the government will determine the amount of reimbursement or equitable adjustment to the contract price.

d. The WSW should clearly state whether redesign is a remedy and under what circumstances the redesign remedy would be invoked. For instance, if the defect is considered systemic, then redesign may be the most appropriate remedy. If redesign requires an engineering change proposal (ECP), then the redesign remedy should state the contractor's responsibility for retrofit.

11. Clause Development. The specific warranty clause including the identification of essential performance requirements (EPR), to be included in the contract must be consistent with the approved WSW Plan. The warranty

clause must be developed by the implementing command program/system manager. Specific requirements that must be addressed in preparing the warranty clause are contained in Attachment 3.

12. WSW Administration. WSW administration requirements must be developed as an integral part of the overall warranty planning and warranty clause development process as required in paragraph 5. Administration requirements must be consistent with the planned operational and maintenance concepts of the weapon system to be fielded, and must be fully integrated with all logistics support elements as defined in AFR 800-8, and any contractor support requirements as defined in AFR 800-21. Additional field level inspections, tests, measurements, or data collection systems must not be required to administer and enforce the terms of the warranty unless these additional requirements are cost-effective, coordinated in the WSW plan, and waivers obtained as required in Attachment 4 for new data systems. The WSW must minimize the administrative burden imposed on maintenance, supply, transportation, and other personnel supporting the weapon system. Specific WSW administration requirements are shown in Attachment 4.
13. Training. Orientation and special training requirements must be established for all personnel responsible for WSW acquisition and administration as described in paragraphs 20 through 23.
14. WSW Management Improvement Group. The WSW Management Improvement Group provides a mechanism for ensuring the timely development and implementation of proposed changes and improvements to WSW policies and procedures. It also recommends the development of other management tools and products, i.e., guide books, models, etc., that contribute to more effective WSWs. This group does not replace or assume any of the responsibilities of the respective major commands or higher headquarters. Rather, this group evaluates and develops recommended changes to WSW policy and implementation tools or techniques. The WSW Management Improvement Group is chaired by SAF/AQCS. The PPAC will serve as executive secretary for the group and will schedule meetings, develop agenda items, and track action items. Other group members will include representatives from SAF/ACCE, HQ USAF/LEYM, HQ USAF/LE-RD, HQ AFSC/PLE/PKC, and HQ AFLC/MMA/PMP. Representatives from other major commands will also be invited to participate to address field concerns and recommended WSW management improvements. The group will meet on a semi-annual basis or more frequently at the call of the chairperson.

Section C—WSW Waivers, Notifications, and Reports

15. Waivers and Notifications Requirements. The policy and procedures for waiving one or more of the WSW guarantees required by 10 U.S.C. 2403, as implemented by DFARS, Subpart 46.770-2, and for notifying the Congress of the Air Force's determination to waive a WSW or not to include EPR guarantees on weapons systems that are major defense acquisition programs not yet in mature full-scale production are contained in DFARS,

Subpart 46.770-9. In addition, the procedures set forth below must be followed.

a. Requests for waivers must be submitted to SAF/AQCS. When contract award would be significantly delayed by the waiver and advance notification, if required, a contract option for the warranty shall be included that can be exercised within a reasonable period of time if the waiver is not approved. If the waiver is being sought on the basis of not being cost-effective, the best price deemed obtainable shall be negotiated between the parties and included in the option.

b. The request for waiver must contain, in addition to the information required by DFARS, Subpart 46.770-9(d), the following:

(1) A copy of the cost-benefit analysis if the basis of the waiver is that a warranty would not be cost-effective.

(2) Action taken to assure product quality and achievement of EPRs in lieu of obtaining a warranty.

(3) Mandatory exercise date of the warranty option, if applicable.

c. Requests for waivers and advanced notifications pursuant to 10 U.S.C. 2403 must be provided to SAF/AQCS, 60 days prior to contract award.

16. Reporting Requirements:

a. WSW Usage Report (RCS: HAF-AQC (SA)(8701)). Each contract award for a weapon system that is covered by a WSW must be reported to SAF/AQCS by the implementing commands. Attachment 5 prescribes the warranty information to be provided. A copy must be sent to AFALC/PP, Wright-Patterson AFB, Ohio. The reports are to be submitted on a semi-annual basis for the periods 1 October to 31 March, and 1 April to 30 September within 45 days after period completion. Within 90 days after each reporting period, the AFALC/PP will provide to SAF/AQCS a summary of the implementing command inputs, including an analysis of such items as warranty duration, exclusion and limitations, unique terms and conditions, etc.

b. Failure Analysis Reports. Contracts containing WSWs must require the contractor to provide failure analysis reports or, as a minimum, corrective action reports for all items returned under the terms of the warranty for corrective action or repair. These reports are distributed to management, engineering, logistics, test and evaluation activities which document a need for such data during the contract data requirements list (CDRL) preparation and to the contract administration office.

c. Incurred Warranty Costs Report. Contracts containing WSW must require the contractor to provide a periodic report of incurred costs as a result of the warranty, if any, to the WSW manager. This report may be in contractor's format and may be submitted as a part of other required cost reports or as a separate report.

d. Warranty Activity Report. The evolving maturity of a weapon system and an adequate performance data base may demonstrate that the continued use of a WSW on future buys is not feasible or cost-effective. Therefore, annual reports by the government or contractor that provide a summary of warranty activity must be accomplished for all contracts containing WSWs. These reports must be submitted to the government program manager. The first report must be provided 1 year from the delivery of the first warranted item under the contract. Subsequent reports must be provided until all item warranties have expired and all claims are settled. The warranty assessments should be used by the program manager to determine warranty provisions and tasks for follow-on procurements for the weapon system, and to evaluate the overall effectiveness of the WSW. It should also be used as a key data input when accomplishing the required CBA that is addressed in paragraph 6. A copy of these reports must be provided to AFALC/PP for information. The report must include as a minimum:

(1) The contractor and contract number.

(2) A summary of the claim activity during the period and cumulative claim activity. Claim activity must include the claims submitted, honored, disputed, and denied, and include the dollar value for each category. Denied claims must include reasons for denials, such as false-pull (not defective), abuse, or not covered by the warranty.

(3) A "remarks" section that identifies the warranted tasks or services that are considered desirable or undesirable based on the claim frequency, failure mode, and dollar value.

Section D—Product Performance Agreements (PPA)

17. PPAs incorporate warranties and other contractual arrangements that motivate the contractor to achieve desired performance. Various incentives may also be included in these arrangements. There are many types of PPAs besides warranties (e.g., reliability improvement warranty, availability guarantee, logistics support cost guarantee, etc.). PPAs provide increased flexibility to tailor the WSW to the program office's needs. Their use should be considered in the EPR portions of the WSW as a means of assuring or providing incentive to exceed performance requirements. The US Air Force PPAC, located at the Air Force Acquisition Logistics Center, Wright-Patterson AFB, Ohio, is chartered to assist acquisition activities in the development and analysis of PPAs. In this regard, they have developed a PPA guide and a decision support system that are readily available. Additionally, they can provide assistance to program offices on any aspect of planning, clause selection, analysis, and administration to develop a WSW that meets program objectives.

Section E—Responsibilities

18. Air Force Secretariat:

a. SAF/AQ:

(1) Establishes Air Force policy on the development, selection, application, implementation, and reporting of warranties in compliance with the regulatory requirements of the FAR, Subpart 46.7 and supplements thereto.

(2) Monitors the WSW Program to ensure implementation is effective and consistent with Air Force and DOD direction.
(3) Reviews requests for warranty waivers and notifications, and forwards such requests to the Assistant Secretary of the Air Force for Acquisition for approval.
(4) Collects, evaluates, coordinates, and submits warranty reporting data to the Secretary of the Air Force, DOD, and the Congress as appropriate.

b. SAF/AC:
(1) Establishes Air Force policy on the selection and application of cost benefit analysis techniques for evaluating alternative warranty strategies as required by FAR, Subpart 46.7 and supplements thereto.
(2) Monitors the WSW program to ensure cost benefit analysis application is effective and consistent with Air Force and DOD direction.

19. HQ USAF/LE:
a. Establishes Air Force policy and guidance in conjunction with SAF/AQ, for the field administration, identification, processing, control, and failure reporting of warranted items in the logistics system.
b. Monitors the WSW Program to ensure implementation is effective and consistent with Air Force and DOD direction.

20. Implementing Command (Usually Air Force System Command, Air Force Logistics Command, or Air Force Communications Command):
a. Designates a command office of primary responsibility for WSW Program policy and implementation.
b. Issues supplemental policy and implementation procedures jointly with the supporting command to fully implement the WSW Program and ensure a smooth transition of warranty management responsibilities during program management responsibility transfer (PMRT).
c. Develops training requirements and implements training programs to ensure that program managers and all responsible program personnel are fully aware of their responsibilities under the WSW Program and that warranty contract requirements are cost-effective, enforceable, and can be administered in the field.
d. Collects, evaluates, coordinates, and submits required warranty data to SAF/AQ for evaluation and further processing to SAF, DOD, and the Congress as appropriate.
e. Designates EPRs in accordance with paragraph 8.
f. Ensures that warranty costs (administration, data, transportation, etc.) are planned and programmed for each weapon system acquisition.
g. Assumes the responsibilities of paragraph 27 when designated as the responsible OT&E organization by HQ USAF Program Management Directive (PMD).
h. Ensures, in conjunction with the supporting command, that warranty contract requirements are cost effective, enforceable, and can be administered in the field.
i. Leads the warranty planning efforts with the participation of the supporting and using commands.

21. Supporting Command (Usually Air Force Logistics Command):
    a. Designates a command office of primary responsibility for WSW Program policy and implementation.
    b. Issues supplemental policy and implementation procedures jointly with the implementing command to fully implement the WSW Program.
    c. Designates a warranty focal point to ensure a smooth transition of warranty management responsibilities during program management responsibility transfer (PMRT).
    d. Develops training requirements and implements training programs to ensure that all support personnel are fully aware of their responsibilities under the WSW Program.
    e. Ensures, in conjunction with the implementing command, that warranty contract requirements are cost-effective, enforceable, and can be administered in the field.
    f. Participates in the warranty planning efforts led by the implementing command, and provides command coordinated EPRs with recommended approaches for administration and tracking of proposed EPR parameters.
    g. Collects, evaluates, coordinates, and submits warranty reporting data as requested by the implementing command, and monitors the effectiveness of procured warranties in achieving WSW Program objectives.
    h. Assumes the responsibilities of paragraph 27 when designated as the responsible operational test and evaluation (OT&E) organization by HQ USAF Program Management Directive (PMD).
22. Using command:
    a. Designates an office of primary responsibility for WSW Program policy and implementation.
    b. Cooperates with the implementing, supporting, and participating commands in developing and implementing WSW Program requirements.
    c. Participates in the warranty planning efforts led by the implementing command, and provides command coordinated EPRs with recommended approaches for administration and tracking of proposed EPR parameters.
    d. Develops training requirements and implements training programs to ensure that all support personnel are fully aware of their responsibilities under the WSW Program.
    e. In conjunction with the implementing command, ensures that warranty contract requirements are cost-effective, enforceable, and can be administered in the field.
    f. When required by a coordinated WSW plan:
        (1) Collects, evaluates, coordinates, and submits warranty reporting data as requested by the implementing command, and monitors the effectiveness of procured warranties in achieving WSW Program objectives.
        (2) Designates the field level warranty action point to coordinate all warranty related data collection, warranty failure reporting, and

warranted item control and distribution requirements with the warranty manager.

g. Assumes the responsibilities of paragraph 27 when designated as the responsible OT&E organization by HQ USAF Program Management Directive (PMD).

23. Air Training Command:

a. Designates a command office of primary responsibility for WSW program policy and implementation.

b. Cooperates with the implementing, supporting, and participating commands in developing and implementing WSW program requirements.

c. Participates in the warranty planning efforts led by the implementing command, and provides coordinated training and training support information required to implement warranty requirements on each acquisition program.

d. Develops training and training support cost information for cost-benefit analysis, tradeoff studies, and other purposes, as necessary.

e. Coordinates with the USAF Product Performance Agreement Center to develop warranty selection, analysis, administration, and enforcement methodology in formal school curricula.

24. Air Force Systems Command (AFSC):

a. Assumes the responsibilities of paragraph 20 when designated as the implementing command.

b. Assumes the responsibilities of paragraph 21 when designated as the supporting command.

c. Develops WSW program implementation procedures in conjunction with AFLC.

d. Provides program direction and with AFLC jointly oversees the operation, staffing, and funding requirements of the USAF Product Performance Agreement Center.

e. Plans for and ensures an orderly transition of warranty management responsibilities to AFLC during program management responsibility transfer.

25. Air Force Logistics Command (AFLC):

a. Assumes the responsibilities of paragraph 20 when designated as the implementing command.

b. Assumes the responsibilities of paragraph 21 when designed as the supporting command.

c. Develops WSW program implementation procedures in conjunction with AFSC.

d. Provides program direction and with AFSC jointly oversees the operation, staffing, and funding requirements of the USAF Product Performance Agreement Center.

e. Ensures an orderly transition of warranty management responsibilities from AFSC during program management responsibility transfer.

26. Air Force Communications Command (AFCC):

a. Assumes the responsibilities of paragraph 20 when designated as the implementing command.

b. Assumes the responsibilities of paragraph 21 when designated as the supporting command.

c. Develops WSW program implementation procedures in conjunction with AFLC.

d. Plans for and ensures an orderly transition of warranty management responsibilities to AFLC during program management responsibility transfer.

27. Air Force Operational Test and Evaluation Center (AFOTEC):

a. Designates an office of primary responsibility for WSW Program policy and implementation.

b. Coordinates the OT&E plan with the implementing, supporting, using, and training commands to ensure that all warranty deficiencies discovered during OT&E are reported as required by the approved warranty plan and Attachment 4 to this regulation.

c. Reports warranty deficiencies, as required, in interim and final OT&E reports.

d. Provides when requested by the implementing command an assessment of the testability of proposed EPRs prior to production contract award.

e. Assists the appropriate organization with the responsibilities in b through d above when such organizations are designated by program management directive (PMD) to conduct the OT&E or when OT&E is initiated and conducted by a major command.

28. USAF Product Performance Agreement Center (PPAC):

a. Develops management tools, analytical techniques, and handbooks to assist program managers in selecting, evaluating, applying, and administering warranties for weapon systems, equipment, and parts.

b. Provides technical assistance, on a consultation basis, to Air Force activities in developing, selecting, and tailoring warranties.

c. Maintains the WSW Program repository and data base to support warranty effectiveness studies and lessons learned requirements.

d. Maintains a repository or locator for warranty-related software developed by the government or at government expense to manage or administer warranties.

e. Develops generic warranty clauses for higher headquarters consideration to minimize the proliferation of unique warranty provisions that require complicated or nonstandard administrative efforts.

f. Serves as the central data repository for warranties and related business arrangements (paragraph 17).

g. Analyzes the effectiveness of existing and proposed warranties and related business arrangements (e.g., award fees, incentives, contractual provisions, etc.).

h. Develops improved warranties and related concepts, as well as methodologies for selecting appropriate and cost-effective warranties.

i. Formulates proposed policy guidance for SAF/AQC consideration concerning the application of warranties to Air Force acquisitions. Serves as the executive secretary to the WSW Management Improvement Group as required by paragraph 14.

29. Implementing Command Program Manager (PM). The following responsibilities supplement and complement those in AFR 800-2. The PM:
    a. Establishes and implements a WSW program as part of the overall acquisition or modification process, as prescribed in this regulation and FAR, Subpart 46.7, as supplemented.
    b. Structures and establishes an effective warranty team to develop and coordinate the program's WSW requirements as required by this regulation.
    c. Ensures that the WSW plan is developed, thoroughly coordinated, and approved as required by this regulation.
    d. Designates the WSW manager and identifies specific functions and responsibilities assigned to the WSW manager. Delegates authority to the WSW manager to carry out WSW program taskings and requirements. As a minimum, the WSW manager must be tasked to:
        (1) Manage and integrate the performance, operational, and support requirements of the using, supporting, and other participating commands in WSW contract development efforts and the planning for administration of the warranted systems.
        (2) Manage and coordinate warranty application, enforcement, and administrative requirements to include warranted item identification, processing, deficiency reporting, data collection, and item disposition.
        (3) Coordinate and resolve disputes concerning warranty program requirements in conjunction with the appropriate contracting or contract administration office, field or depot action points, legal offices, etc.
        (4) Inform the PM of WSW program status and problem areas requiring special attention or support from higher headquarters or other participating commands.
        (5) Coordinate planning for and ensure a smooth transition of warranty management responsibility transfer as a part of the PMRT planning and implementation effort.
        (6) Provide a copy of the approved warranty plan to the USAF PPAC.

By Order of the Secretary of the Air Force

Official

LARRY D. WELCH, General, USAF
Chief of Staff

WILLIAM O. NATIONS, Colonel, USAF
Director of Information Management and Administration

AFR70-11 Attachment 1
1 December 1988

WSWs and the System Life Cycle

The warranty plan must be approved within 6 months of the award of the FSD contract and updated as appropriate for the initial and follow-on production contracts.

AFR 70-11 Attachment 2
1 December 1988

WSW Plan Requirements

Warranty plans must be developed by the implementing command program office and coordinated with using and supporting commands, as well as the cognizant contract administration office and other organizations which are tasked in the plan for WSW support. The program manager has overall responsibility for warranty planning and the establishment of the warranty team to prepare and coordinate the plan.

The warranty plan must be approved by the program manager within 6 months after FSD contract award, and updated as needed to provide warranty implementation requirements for fielding the warranted item. The warranty plan must also be updated to reflect any change in requirements prior to the award of follow-on production contracts.

The warranty plan must address the following:

1. *Acquisition Background.* Describe the weapon system being acquired. Summarize the program and warranty history to date, including an explanation of why DFARS, Subpart 46.770, applies.
2. *WSW Clause.* Attach the proposed warranty clause to the plan and identify here any special considerations or constraints affecting selection of the terms and conditions. The clause must address the requirements of Attachment 3 to this regulation or rationale provided in the WSW plan for the exclusion of any of those requirement(s).
3. *Cost-Benefit Analysis (CBA).* Describe the CBA methodology used and summarize the CBA results.
4. *Warranty Administration.* Describe the specific requirements (e.g., markings) to administer the warranty as identified by Attachment 4 to this regulation. Ensure that the administrative requirements of the proposed warranty clause are consistent with this section of the plan.
5. *Warranty Team Membership.* Describe the warranty team organizational and management responsibilities. List the team membership (e.g., warranty manager, contracting officers, engineers, logisticians, cost analysts, using and supporting command representatives, and other points of contact deemed necessary for warranty administration).

6. *Program Management Responsibility Transfer* (*PMRT*). When applicable, identify the planned approach to transition warranty enforcement and administration responsibilities from the implementing command to the supporting command and summarize the CBA results at this point in the program.
7. *Foreign Military Sales* (*FMS*). If a WSW is to be obtained for an FMS purchaser, the FMS purchaser's warranty requirements and the Air Force plan to obtain those requirements should be discussed here. A separate FMS warranty plan may be developed if the FMS purchaser has requested unique warranty requirements that dictate the need for more detailed planning.
8. *Contractor Support.* If contractor support (e.g., contractor logistics support (CLS) or interim contractor support (ICS) is planned, ensure that the support requirements are clearly defined, compatible with the WSW, and the related costs of each (i.e., the WSW and ICS or CLS) are segregated for accounting purposes.
9. *Schedule.* Identify key events and dates such as delivery dates, warranty periods, CBA accomplishment and updates, etc.

AFR70-11 Attachment 3
1 December 1988

WSW Clause Development

The terms and conditions of the WSW must be tailored to the weapon system and must be as clear and simple as possible with emphasis on enforcement of the warranty conditions through existing Air Force management, administration, and logistics processes. The following requirements must be included in the warranty terms and conditions unless the warranty plan provides rationale for the exclusion of the requirement and approval has been granted, if required:

1. Define key terms such as, acceptance, defect, correction, remedy, etc.
2. Incorporate the three guarantees required by Title 10, U.S.C., Section 2403, as addressed in DFARS, Subpart 46.770-2, and paragraph 4b of this regulation, unless a waiver is granted.
3. Describe the roles and responsibilities of the government and contractor in the warranty process.
4. Identify the production units covered by each of the three parts of the warranty and the units, if any, excluded from the warranty coverage.
5. To the maximum extent possible, state the warranty duration as a fixed period of time from date of delivery. WSW duration must be of sufficient length to determine that the WSW requirements have been achieved. When the duration is based on item utilization rather than calendar time, appropriate measuring devices or techniques (e.g., elapsed time indicator, cycle counter) must be required. Warranty duration should allow for those

anticipated non-operational activities after delivery such as, transportation, storage or shelf-life, and redistribution. Other warranty duration considerations that should be addressed are whether:

a. Warranty duration applies to an individual unit or to a group or subgroup.

b. Warranty duration starts with acceptance (delivery) or at time of installation of the unit in a higher level of assembly.

c. Warranty period can be extended and under what conditions (e.g., to compensate for warranty time lost while a defective unit was being repaired or replaced).

6. Describe the EPRs to be warranted. Include a description of how they are to be measured, when they are to be verified, and any special testing and test equipment required to complete the verification. Also identify the contractor's role and responsibility in the verification.
7. All warranted items must be marked in accordance with MIL-STD-129, Marking for Shipments and Storage, and MIL-STD-130, Identification Marking of US Military Property, except for items which cannot be effectively marked. Markings must be located in a manner so as to be conspicuous to the person removing the item from service, and the period or conditions of the warranty must be specifically stated (e.g., landings, flight hours, operating hours, days from shipping, date of expiration, etc.).
8. Describe the remedies available to the government if the contractor breaches the WSW. Conditions for invoking a particular remedy should be addressed. When contractor repair is stipulated as an authorized remedy, also stipulate the required turn around time from contractor receipt of the failed item to contractor shipment or government acceptance of the repaired or replacement serviceable item. Also indicate the government remedy should the contractor fail to meet the guaranteed turn around time. If government repair of the hardware or associated software is to be authorized as a part of the stipulated remedies, clearly identify the conditions, limitations, and exclusions that apply as well as the repair rates at which the contractor will reimburse the government for government repaired items. Also, the WSW must state whether redesign is a remedy and under what circumstances it would be invoked.
9. Describe all warranty data and report requirements and include appropriate contract data requirements list (CDRL) for distribution to the cognizant contracting, engineering, logistics, and test activities.
10. Address the impact on the warranty should the government determine to break out any of the weapon system's component parts or use other qualified spare parts in the repair of the warranted system.
11. Identify any exclusions such as, mishandling, fire, combat damage, etc., (see DFARS, Subpart 46.770-3, as supplemented).
12. Identify any limitations such as the contractor's financial liability (see DFARS, Subpart 46.770-3, as supplemented).
13. Establish warranty terms and conditions consistent with the weapon system's operational and maintenance concept and the warranty administration requirements in Attachment 4. Do not require additional government inspections, measurements, data collection, or other unique administra-

tive processes to enforce the warranty unless demonstrated to be cost effective in the CBA, coordinated in the WSW plan, and a waiver was obtained as required in Attachment 4 for new data systems.

14. Include a statement that the warranty does not limit the government's rights under any other contract clause.
15. Establish packaging and handling requirements for warranted items according to the level of protection as specified in MIL-STD-2073-1A or as specified in a government approved special packaging instruction. Packaging and handling costs are not directly reimbursable to the government, but should be considered in the remedy for correction of failed warranted items.
16. Establish transportation requirements after obtaining traffic management office (TMO) advice as required by DFARS, Subpart 47.101. The program office and TMO must consider the following in developing transportation requirements:
    a. The government arranges and bears actual transportation costs of US Government-owned assets that are returned to the contractor for correction or replacement of defective or non-conforming parts, and the contractor reimburses the government at a pre-negotiated reduction in contract price for each return of a failed or defected warranted item.
    b. Assets are to be shipped on Second Destination Transportation (SDT) funds via a mode that will ensure delivery to the final destination within the timeframes of the Uniform Material Movement and Issue Priority System (UM-MIPS).
    c. Transportation costs incurred for the movement of foreign military sales assets to or from the contractor for correction or replacement of defective or nonconforming parts are charged to the foreign country.
17. Describe the process for determining the impact on the WSW of approving a waiver or deviation to a requirement in the contract specification and for determining an equitable adjustment, if any to contract price.
18. Address the prime contractor's responsibility for warranting any property furnished to the contractor as government-furnished property (GFP) (see DFARS, Subpart 46.770-5).

AFR70-11 Attachment 4
1 December 1988

WSW Administration Requirements

1. The warranty administration process begins with contract award and ends when all item warranties have expired and all warranty claims have been settled. Warranty administration requirements must be established to ensure that warranted items are properly identified and defective warranted items are reported, controlled, and corrected under the terms of the contract. A WSW normally includes coverage for essential performance re-

quirements (EPR), defects in materials and workmanship, and nonconformance to design and manufacturing requirements.

2. The following general management requirements must be established to ensure an effective warranty administration program:
   a. Organization:
      (1) A warranty manager is designated by the implementing command program manager to administer, coordinate, and control the administration of warranted systems. The warranty manager coordinates with the appropriate government and contractor organizations to resolve warranty claims and correct warranted deficiencies. Warranty management responsibility may transfer or be assigned to the supporting command as agreed to in the warranty plan and PMRT agreement. When warranty management responsibility transfers to the supporting command, the supporting command must designate a WSW manager to administer the remaining term of the WSW.
      (2) Field and depot organizations operating and maintaining warranted weapon systems designate warranty action points to coordinate the delivery of weapon system performance data and the identification of deficient warranted items to the warranty manager. These organizations must identify defective items covered by the warranty and initiate deficiency reports as required below. Field and depot organizations should not attempt to repair the warranted item unless government repair is authorized under the terms of the warranty.
   b. Warranty Administration Management Systems. Program offices must establish manual or automated management systems to administer WSWs. These systems must be capable of accepting weapon system field performance data to determine whether or not EPRs are achieved. They must also be capable of tracking defective warranted items to ensure that deficiencies are corrected according to the contract remedies. Program offices should contact the USAF Product Performance Agreement Center (PPAC) to locate available government developed warranty management systems. Government costs to develop new or modified automated systems must be factored into the WSW CBA.
   c. Field Level Warranty Information and Orientation:
      (1) General Warranty Information. The USAF Technical Order System may be used to provide field and depot support organizations general information concerning the contractual warranty requirements. Operational supplements, technical order (change) page supplements (TOPS), or supplemental manuals to the appropriate system or item technical order(s) as agreed to in the WSW plan should describe the general warranty parameters and how they are to be measured or tracked, describe or provide a picture of standard warranty markings and indicate where they are normally located on the item, indicate the authorized level of repair for warranted items, and identify those warranted systems or items subject to deficiency reporting in accordance with TO 00-35D-54.

Work unit code (WUC) manuals may also be used to identify those systems or subsystems subject to a warranty. Warranty coded WUCs provide a basis to identify field removal or repair actions on these systems or subsystems. The WSW plan must specify how the TO System is to be used to provide warranty administration information.

(2) Warranty Orientation and Training. For newly fielded weapon systems, a warranty orientation and training program must be established for all personnel who will have responsibilities for administering warranted weapon systems. This orientation may be contractor conducted (but closely monitored by the government) or included as part of government provided special training programs. This orientation and training should be based on the approved WSW Plan as updated to reflect field implementation requirements. The program manager, in conjunction with the using command(s), training commands, and supporting command must develop source materials or contractor requirements for this effort.

3. Administrative Requirements for the EPR Warranty.

a. The WSW EPR coverage requires the collection and evaluation of weapon system performance data against specified contract performance parameters. If the weapon system fails to achieve the specified EPR, then a remedy is due to the government. Normally, this includes contractor repair, replacement, or redesign of subsystems or parts which failed and thereby caused the weapon system to fall short of its warranted EPR. Exact identification of those parts subject to no cost contractor repair, versus those which failed within EPR parameter is not accomplished at the field level. Rather, the program office or warranty manager must:

(1) Establish procedures to ensure that failed parts are turned in to supply and shipped to the contractor for repair or replacement.

(2) Identify those failures related to the EPR value that are to be repaired by the contractor at no cost to the government, and identify those failures for which the government will bear the cost of repair.

b. EPR warranties must be designed so that EPRs can be measured by standard Air Force operational and maintenance data collection systems such as the Core Automated Maintenance System (CAMS), Reliability and Maintainability Information System (REMIS), Comprehensive Engine Management System (CEMS), or Combat Ammunition System (CAS). Elapsed time indicators (ETI) and integrated flight data recording devices may be used to provide associated performance data as appropriate (see AFR 66-6 for the application and use of ETIs). Changes to the above automated maintenance data collection systems solely to accommodate warranty performance data collection must be approved by HQ USAF/LEY. Specialized, automated, or weapon system unique field data collection systems will not be developed or implemented for warranty performance measurement without the prior approval of HQ USAF/LEY.

c. Requirements for weapon system performance data must be identified in the WSW Plan and thoroughly coordinated with the using and supporting commands prior to production contract award. Evaluation of performance data to determine whether the warranted system meets the EPR must be accomplished by the WSW manager with assistance from the using and supporting command as agreed to in the WSW Plan.

d. If the remedy for failure to meet a specified EPR includes contractor replacement and repair of failed parts, the WSW manager must be able to determine which parts are subject to no cost contractor repair. In addition, if government repair is authorized for these parts, then the program office must incorporate negotiated repair rates in the WSW clause at which the contractor will reimburse the government for government repaired items. Data required to identify specific failures and repair actions should be collected by the review of:

(1) AFTO Forms 349 (Maintenance Data Collection Record) or similar CAMS input records with warranty suffix (W) in work unit code block.

(2) Contractor corrective action or failure analysis reports.

(3) Warranty coded service reports or deficiency reports.

NOTE: Field activities will not submit warranty coded service reports or deficiency reports for failed parts to evaluate EPR performance unless this approach is shown to be the most cost effective for the government and the using command agrees to this requirement in the WSW plan.

(4) Other reports as identified and agreed to in the approved WSW Plan.

4. Administrative Requirements for Warranties Covering Defects in Materiel and Workmanship and Nonconformity to Design and Manufacturing Requirements:

a. General. As indicated above, the failure of a weapon system to meets its EPR requirements is identified through the evaluation of weapon system performance data. However, defects in materiel and workmanship and nonconformance to design and manufacturing requirements are identified as a result of personal inspection and evaluation of a part in its intended use. Normally these types of engineering, manufacturing, or quality deficiencies are discovered as a result of a system or part failure. Warranted weapon systems with defects in materials and workmanship and nonconformance to design and manufacturing requirements must be identified, reported, and processed as indicated below.

b. Failure Reporting and Processing Requirements for Defective Warranted Items:

(1) Warranty-coded service reports (W-SR) or warranty-coded deficiency reports (W-DR) must be prepared and submitted for defective or nonconforming warranted items when required by the warranty instructions contained in the system or component level technical order. A W-SR or W-DR is normally a Service Report (SR), Material Deficiency Report (MDR), Quality Deficiency Report

(QDR), or Software Deficiency Report (SDR) with a warranty "yes" indicated in block 19 of the report. If a SR, MDR, QDR, or SDR is not required for a failed part, then a separate warranty deficiency report (WDR) is prepared. The W-SR, W-DR, or WDR is prepared according to TO 00-35D-54. Generally, the W-SR is prepared for those acquisition programs managed by the implementing command prior to PMRT or those systems undergoing test and evaluation as defined by AFR 80-14. After PMRT, W-DRs and WDRs are submitted to the supporting command (AFLC). The W-SR, W-DR, or WDR is prepared by those field and depot organizations responsible for operating and maintaining warranted weapon systems upon discovery of failed or nonconforming warranted items and prior to the turn-in of the failed item.

(2) Action points must determine whether further examination and investigation are required; i.e, service reporting or materiel deficiency reporting, over and above the failure analysis and reporting requirements of the warranty in accordance with TO 00-35D-54. If such investigation is required, the processing and disposition of the W-SR, W-DR, or WDR and the warranted item (exhibit) must be coordinated with the warranty manager as outlined in TO-00-35D-54.

(3) If a weapon system, subsystem, or part is not identified as warranted for defects in materials and workmanship and nonconformance to design and manufacturing requirements as required by this regulation, i.e., labels, work unit codes, etc., then field activities may assume that no warranty exists and, therefore, no W-SR or WDR is required. If an item is properly marked as warranted, and the warranty instructions in the applicable technical order supplement direct submission of W-SRs, W-DRs, or WDRs, a W-SR, W-DR, or WDR must then be submitted to the action point as required by TO-00-35D-54. The action point provides copies of W-SRs, W-DRs, and WDRs to the warranty manager and coordinates all action with the warranty manager.

(4) Upon receipt of the W-SR, W-DR, or WDR, the WSW manager completes an investigation as part of the warranty corrective action and provides disposition instructions through the action point to the originating and screening points. Normally, predisposition instructions will be provided to the originating and screening points to avoid delays and to expedite the warranted item processing.

(5) Defective warranted items are controlled, e.g., handled, received, stored, shipped, and processed in accordance with TO 00-35D-54; AFM 67-1, Volume 1, Part One, Chapter 10, section J; and AFM 67-1, Volume II, Part Two, Chapters 10 (Receipt Processing), 11 (Issue Systems), and 15 (Shipment); as appropriate. Close coordination with the action point and warranty manager is required to ensure timely processing, proper identification and storage, proper packaging and transportation, and other administrative requirements are completed to ensure that the full benefit of the warranty is obtained by the government.

(6) Warranted items, i.e., covered by materials and workmanship and design and manufacturing warranties, must be marked according to MIL-STD-130 and item containers must be marked according to MIL-STD-129. Warranty items must be packaged according to the requirements of the original contract, MIL-STD-2073-1A, or any Air Force special packaging instruction per AFR 71-9. For defective warranty items, the level of preservation and packaging must be those specified for unserviceable condition items.

c. Warranty Corrective Action. The warranty manager notifies the appropriate government and contractor activities, in coordination with the appropriate contracting officer, that a defective warranted item has been identified and that corrective action or remedy under the terms of the warranty is required. Corrective action or remedy may include: return and no cost repair or replacement of the item by the contractor, repair by government activity with defective component(s) returned to the contractor for no cost repair or replacement, repair by government activity with compensation or contract price reduction, or other remedies as provided for in the warranty terms and conditions.

Note: In the event that operational mission requirements preclude executing the corrective action or remedy under the terms of the warranty, the action point must document the circumstances and rationale for the action and provide a written notification of such action to the warranty manager. This action must be approved in advance by the unit commander or his or her designated responsible officer. This approval may not be delegated lower than the Deputy for Maintenance, Deputy for Resource Management, or equivalent level officer.

d. Warranted Item Accountability. Accountability for defective warranted items which are returned to the contractor for repair, replacement, or investigation must be maintained by the government. WSW managers, in conjunction with the appropriate contract administration office, must maintain cognizance over warranted items shipped from government installations to contractors facilities for repair or replacement. Repaired warranted items must be returned to the government within the time frame established by the warranty and with the proper markings to indicate the new warranty period of performance.

e. Commercial Item Warranties. Commercial off-the-shelf items that are an integral part or subsystem to the weapon system being procured for which a warranty is required, must be identified, controlled, and administered under the warranty provisions in the weapon system contract and as indicated herein. Warranty administration procedures contained herein should be used to the maximum extent possible for all other commercial off-the-shelf items which have a standard commercial warranty.

AFR70-11 Attachment 5
1 December 1988

WSW Usage Report Format (RCS:HAF-AQC(SA)(8701)

Part I—Weapon System Warranties

For each production contract for a weapon system that includes the warranty required by Title 10, U.S.C., Section 2403, as implemented by DFARS, Subpart 46.770-2, the following data is required:

A. System Nomenclature: (F-15, GBU-15, Peacekeeper, etc.)
B. Warranty Scope: (System/subsystem covered by the warranty, such as Inertial Navigation Unit, Engine, etc.)
C. Contract Number:
D. Contractor:
E. Contract Value: (Including priced options.)
F. System/subsystem Quantity: (Total number of warranted systems or subsystems.)
G. Warranty Costs: (As set forth in an applicable contract line item, if separately priced; or as reflected in the government price negotiation memorandum, if estimated.)
H. Warranty Cost as a Percent of Contract Value: (Paragraph G. divided by E.)
I. Warranty Cost Cap (if any): (A cap negotiated to limit the financial liability of the contractor to correct defects.)
J. Contract Environment: (Competitive, non-competitive, etc.)
K. Warranty Provision(s): (FAR, DFARS, special provision.)
L. Warranty Coverage: (Materials and workmanship: design and manufacturing: and/or EPR, such as, functional reliability, maintainability, availability, etc.)
M. Warranty Duration: (Calendar time; operating/flying hours; etc.)

Part II—Summary.

This part provides an overall summary of WSW information provided under Part I.

Total No. Contracts	Total Value of Contracts	Total Warranty Costs	Total Warranty Costs as % of Total Value of Contracts
(a)	(b)	(c)	(c/b)

Appendix

G

Marine Corps Order 4105.2

DEPARTMENT OF THE NAVY
HEADQUARTERS UNITED STATES MARINE CORPS
WASHINGTON, D.C. 20380-0001

MCO 4105.2
LMA-4-DT
4 Nov 1987

MARINE CORPS ORDER 4105.2

From: Commandant of the Marine Corps

To: Distribution List

Subj: Marine Corps Warranty Program

Ref:
(a) Public Law 98-525, Defense Procurement Reform Act of 1985 (NOTAL)
(b) Defense Federal Acquisition Regulation Supplement (DFARS) 46.7 (NOTAL)
(c) Navy Acquisition Regulation Supplement (NARSUP) 46.7 (NOTAL)
(d) DoD-Hdbk-276-1
(e) MIL-STD 881A
(f) MCO 4855-10A
(g) MCO P4000.21A
(h) MIL-STD 130
(i) MIL-STD 129
(j) NavCompt Manual, volume 4, 043108

Encl:
(1) Definitions
(2) Standard Warranty Procedures
(3) Expected Failure Concept
(4) Warranty Claim Data Report Format

Report Required: Warranty Claim Data (Report Symbol MC-4105-01), par. 5b(9), and encl (4)

1. *Purpose.* To promulgate policy described in references (a) through (c) and assign responsibilities for the management and execution of the Marine Corps Warranty Program.
2. *Background.* Reference (a) added Section 2403 to Title 10 of the United States Code and requires the Department of Defense (DoD) to obtain warranties in contracts for weapon systems awarded after 1 January 1985. Specifically, the section requires that weapon systems with a unit cost of more than $100,000 or a projected total procurement cost of more than $10,000,000 possess a warranty in which the contractor warrants:
   a. That the weapon system conforms to the design and manufacturing requirements specifically cited in the contract.
   b. That the weapon system is free from defects in material and workmanship.
   c. That the weapon system meets or exceeds the essential performance characteristics specifically delineated in the contract.

   This section describes various remedies for the contracting officer should the warranty be invoked. These include: requiring the contractor to promptly take action to correct the deficiency at no additional cost to the Government or requiring the contractor to pay costs incurred by the Government to correct the problem. The law requires contracting officers to tailor warranties to fit the particular acquisition and describes criteria for waiving warranty requirements on systems acquisitions. These areas are discussed in greater detail in the following sections.
3. *Objectives.* The objectives of the Marine Corps Warranty Program are to ensure that the weapon systems acquired perform as required, conform to the design and manufacturing requirements specified, are free from defects in materials and workmanship, and finally, to ensure that the new weapon systems/equipment contribute to increased readiness throughout the Marine Corps.
4. *Policy.* The stated objectives can best be accomplished through the judicious development, acquisition, and implementation of performance assurance warranties for new weapon systems and selected equipment. The following policy is applicable for all acquisitions in which the Marine Corps is the contracting authority:
   a. A warranty shall not be used as a substitute for proper logistics planning and acquisition of the elements of integrated logistics support for the system or as a means of acquiring interim contractor support. Warranty considerations shall become part of the acquisition planning process and acquisition documentation.
   b. Per references (b) and (c) the Marine Corps shall acquire only those warranties demonstrated to be cost-effective. A documented cost benefit analysis shall be used to determine the cost effectiveness of a proposed warranty. Prior to performing the analysis, the Government shall require the contractor to identify all contractor costs associated with the warranty or to separately price the proposed warranty. The analysis shall become part of the contract file and program documentation.
      (1) The cost benefit analysis shall be a comparison of the life cycle costs without a warranty and the life cycle costs with a warranty. The

warranty cost benefit (WCB) shall be defined as the result obtained when subtracting the life cycle costs with a warranty from the life cycle costs without a warranty. If the WCB is equal to zero or is positive it may be assumed that the warranty is cost-effective. If the WCB is negative, then the warranty may be assumed not to be cost-effective and a waiver should be requested using the procedures described in paragraph 4c, following. As a minimum, the following cost factors shall be included in the life cycle cost computation for the analysis:

(*a*) Estimated cost to the Government (price) of the warranty.

(*b*) Estimated cost for correction or replacement by the Marine Corps.

(*c*) Estimated cost for correction or replacement by another source.

(*d*) Indirect costs incurred by the Marine Corps to maintain the warranty in effect. Examples of indirect costs include, but are not limited to; costs of warranty defaults, reduced opportunities for breakout, and reduced opportunities for competition.

(*e*) All administrative costs associated with tracking and processing warranty claims, maintaining warranty related records, and reporting of warranty related information. (*Note:* reference (d) can be used to perform the analysis in the detail necessary.)

c. When determined to be cost-effective, the Marine Corps shall acquire warranties on weapon systems/equipment that meet the following criteria, unless a waiver of the warranty requirement has been approved by the Assistant Secretary of the Navy (Shipbuilding and Logistics) (ASN(S&L)). A warranty is required if the system or equipment:

(1) Is a weapon system, as defined in enclosure (1), with a unit cost exceeding $100,000 or with a projected total procurement cost exceeding $10,000,000.

(2) Is an item subordinate to the weapon system level and:

(*a*) Falls within the cost criteria described in paragraph 4c(1). These items would include major components of the system or other equipment integrated to form a system. Spare parts will not be subject to warranties under this Order.

(*b*) Occurs no lower than level 3 of the work breakdown structure of the system. (Refer to reference (e)).

(*c*) Is not reparable at a level lower than fourth echelon.

d. When the Principal Development Activity (PDA) or contracting authority is other than the Marine Corps, the Marine Corps shall provide its warranty requirements to the PDA or contracting authority for inclusion in the contract. In the event the other service PDA or contracting authority has developed warranty provisions for the proposed contract the Marine Corps shall acquire that warranty as long as it does not violate the policy described herein. The policy described herein is not applicable to weapon systems/equipment procured and supported totally through Navy appropriations (i.e., aviation weapon systems and equipment) but may be used for guidance in structuring warranty provisions for those systems/equipment.

e. The Marine Corps shall tailor warranties, consistent with the requirements of this Order, to meet the unique circumstances of each acquisition. Warranties acquired by the Marine Corps shall generally provide for two types of coverage; these are, systemic defect and individual item failure coverage.

(1) Systemic defect coverage provides coverage for the entire weapon system. This level of coverage is appropriate when describing essential performance characteristics for the system. Indicators of systemic deficiencies are frequent Quality Deficiency Reports (QDR) on particular parts of the system or the system itself that establish a trend of failures indicating a possible design deficiency as well as the inability of the system to meet the contractually specified essential performance characteristics. When systemic defects exist, the warranty remedy should call for total asset remedies which could take the form of recalls, repairs, contract price reductions, or combinations of these.

(2) Individual item coverage refers to the coverage extended to those components reparable at the 4th echelon or higher and those warranted parts occurring at/or above level 3 of the work breakdown structure. Items selected for individual item warranties should normally be high dollar components.

f. Warranties shall be acquired for equipment that does not meet the definition of a weapon system only when they are demonstrated to be cost-effective.

g. Commercial warranties are often available when procuring nondevelopmental items. In cases where a commercial warranty is available it will normally be acquired instead of negotiating a separate warranty agreement. These warranties may be acquired if one of the following is true:

(1) They are cost-effective and can be executed with existing supply and maintenance procedures to include administrative procedures for tracking and executing the warranty.

(2) The warranty cost cannot be severed from the item price to effect a price reduction for the item.

**h.* The Marine Corps shall seek a waiver from the ASN(S&L) when the results of the cost/benefit analysis indicate the acquisition of a warranty would not be cost-effective; when relief is desired from one of the three areas requiring warranty coverage as described in paragraph 4j, following, or when it is in the interest of the national defense not to have a warranty on a particular system. These waivers shall be initiated by the Program Manager (PM) and processed, via the Commanding General (CG), Marine Corps Research, Development, and Acquisition Command (MCRDAC), and forwarded for approval to the ASN(S&L).

i. Acquired warranties shall be compatible with existing Marine Corps supply and maintenance procedures so that support for the item, while under warranty, will not differ from the follow-on support provided after the warranty expires. Using unit participation in the implementation, execution, and administration of warranties shall be kept to a minimum; this includes minimizing the imposition of additional supply and maintenance administrative procedures for tracking and administering warranties for equipment in using units (i.e., first, second, and third echelon maintenance capable units). The following shall be considered when developing warranty terms:

(1) The requirements for storage or service of warranted items, while under warranty, shall not differ from their post warranty requirements.

(2) Supply support procedures for warranted items shall operate within the existing Marine Corps supply system.

(3) Existing Marine Corps maintenance management procedures shall be used to document maintenance on warranted items. Warranty claims shall be submitted by the warranty coordinators per enclosure (2) and the provisions of reference (f).

j. The Marine Corps shall require the contractor to warrant that the weapon system provided under the contract conforms to the design and manufacturing requirements specified in the contract; the weapon system provided under the contract is free from all defects in materials and workmanship; and the weapon system, if manufactured in mature production, conforms to the essential performance characteristics specified in the contract. References (a), (b), and (c) apply.

k. Contracts with a warranty shall contain terms that permit the contracting officer to require the contractor to take whatever corrective action is necessary at no cost to the Government to correct the deficiency, to equitably reduce the contract price, or to require the contractor to pay costs reasonably incurred by the Government to correct the deficiency. Corrective action shall be completed within time limits specified in the contract. Contract terms shall allow the Marine Corps to repair a warranted item.

l. The Marine Corps shall not require a contractor to warrant government-furnished equipment (GFE).

m. The Marine Corps shall not seek warranties in cost reimbursement contracts.

n. The Marine Corps shall seek warranties on technical data as defined in reference (g) when it is cost-effective to do so. Warranties for technical data should be subjected to the same criteria and cost effectiveness requirements as their hardware counterparts. Warranted technical data shall be marked to indicate the warranty coverage and expiration date.

o. The Marine Corps shall use the expected failure concept detailed in enclosure (3) when developing the item warranty.

p. The duration of the warranty should be of sufficient time to ensure that those items placed in storage will have warranty protection upon placement in service. In some cases, when extended storage (storage duration to exceed 1 year) is planned for new equipment (i.e., Selected Marine Corps Reserve (SMCR) units) the Marine Corps may seek provisions in the contract that provide for extended warranty coverage for equipment placed in extended storage.

(1) The warranty duration shall be expressed in two terms, the first being some measure of operational use such as miles, hours of opera-

tion, rounds fired, etc. which is sufficient in quantity to ensure the quality of the system/equipment. The second term shall be a period of time extending from the date of acceptance for a period of days, months, or years into the future during which the Government may seek remedies as defined in the contract to deficiencies in the system. For example, a new truck is being fielded with several vehicles in the first production lot destined for delivery to the Maritime Prepositioned Ship (MPS) Program for at least 1 year of storage. Average yearly mileage for the truck is 12,000 miles and at least 1 year of warranty protection is desired for all operational vehicles. The warranty duration might read, "12,000 miles or 24 months whichever comes first." This would permit storage of some vehicles for up to 1 year and still allow for warranty coverage when put into operation. Those vehicles immediately put into operation would have up to 24 months of warranty protection as long as their mileage remained under the 12,000-mile limit.

(2) The warranties on individual items within a warranted system shall not have durations beyond that of the system warranty. In addition, if a warranted item is replaced prior to the expiration of the system warranty, the remaining duration on the individual item warranty shall not exceed the duration remaining on the system warranty. For example, a truck with a 24-month warranty on the entire vehicle also has individual item coverage for the engine for a period of 24 months. The engine fails and is replaced 18 months into its life. The new engine only has 6 months of warranty coverage remaining.

q. Items covered under a warranty shall be marked with the following information at a minimum: "WARRANTY ITEM," production contract number, production lot number, and expiration date/usage factor for the warranty for that production lot. For further information on the marking of warranted items, refer to references (h) and (i).

r. The Marine Corps shall tailor warranties to meet the unique circumstances of each acquisition.

(1) During the tailoring process the Marine Corps shall seek to limit systemic coverage to between three and seven essential performance characteristics; one of which shall be a system level reliability value accompanied by clear definitions of system failures.

(2) Individual item warranty coverage for parts of the system shall be limited to those items reparable at fourth or fifth echelon maintenance or appear no lower than level 3 of the work breakdown structure.

(3) During the tailoring process the Marine Corps shall use the expected failure concept described in enclosure (3) as the principal means of structuring the warranty. In some commercial warranties tailoring may not be possible. In these cases a failure-free warranty may be more appropriate.

s. The Marine Corps shall collect information on the use of warranties for analysis and reporting. This information shall include identification of the contract, the contractor, a summary of claim activity for the reporting period, and the cumulative claim activity for the contract. Claim activity shall include claims submitted, honored, disputed, denied, and the value of each category. Denied claims shall include the reason for denial and failure cause, if known.

t. The Marine Corps shall ensure that one or more of the following remedies are available to the Government when a warranty is breached for a weapon system or equipment. These remedies shall be clearly described in the provisions of the contract.

(1) In cases where a production contract is still in place, the Marine Corps shall seek to equitably reduce the contract price in an amount equal to the cost of parts, transportation costs, handling costs, and labor costs if any are involved. For ease of administration, the reductions should be accomplished in block modifications to the contract on a quarterly or semi-annual basis as specified in the contract.

(2) If a production contract is not in place, the Marine Corps shall seek replacement of the faulty parts or components covered by the warranty. If labor costs are involved, the contracting officer shall seek additional spares in type to the ones that failed and in an amount equal to the value of the labor costs incurred in lieu of receiving monetary reimbursement for the items. Transportation and handling costs for replacing warranted items shall be borne by the contractor.

(3) Monetary reimbursements for parts, labor, and other costs shall be considered to represent a reduction in the contract price or an overpayment to the contractor. As such, the proceeds shall be collected and accounted for using procedures described in reference (j). These proceeds shall revert to the appropriation or appropriations concerned when the issue of reimbursement is covered by an agreement between the contracting parties. Monetary reimbursements shall be addressed in the warranty provisions of the contract.

(4) When the system fails to meet its essential performance characteristics as evidenced by a trend analysis of QDRs, by the systems failure to perform as required, or when the number of system failures exceeds the threshold established for the system, the Marine Corps/Government agent shall seek redesign of the component, subsystems, or system (as necessary) to ensure the system conforms to the essential performance characteristics described in the contract. Additionally, the contractor shall be required to bear the costs of modifying existing inventory (end items and spare parts) to correct the deficiency. Such redesign, testing, modification, and related costs shall be borne by the contractor. Provisions for warranty coverage of redesigned components, subsystems, or system should be described in the contract.

(5) Warranty remedies shall not be any less responsive than normal Marine Corps supply and maintenance turnaround times. Responsiveness in terms of time between the warranty claim notification and resolution shall be addressed in the warranty provisions of the contract.

u. Warranty procedures shall allow for the Marine Corps to effect its own repairs without voiding the remaining warranty. The cost of Marine Corps repair of a defect which is covered by the warranty shall be at the contractor's expense.

v. Warranty procedures identified in enclosure (2) shall be tailored to the designated equipment and included in an advance logistics order (ALO).

5. *Responsibilities*

a. The CMC is responsible for the following:

(1) The CMC (L) shall:

*(a) Issue policy for the technical and statutory requirements of warranties for Marine Corps acquired items (L).

*(b) Issue policy regarding data collection and reporting used to identify warranties, determine compliance, and ensure that acquired warranties are compatible with standard Marine Corps supply and maintenance procedures (L).

**b.* The CG MCRDAC shall:

*(1) Assist PMs in the preparation and tailoring of warranty provisions for systems and equipment (PSI-L).

*(2) Review and forward for approval to the ASN (S&L) all requests for waivers to the warranty requirements identified in reference (b) and maintain copies of all requests as part of the program documentation.

*(3) Review trend analyses of QDRs submitted by Marine Corps Logistics Base (MCLB), Albany, to determine if essential performance characteristics are being met or a design deficiency exists (PSI-G).

*(4) Develop policy for the technical and statutory requirements of warranties, data collection, and compliance determination for the warranty program (PSI-L).

*(5) The PM shall:

(*a*) Identify within an ALO the following information: the essential performance characteristics included in the warranty, the national stock number (NSN) of individual warranted items, the duration of the warranties, and a description of the warranties at the system or individual item level. Identify any procedures that deviate from existing ones in implementing, executing, reporting, or administering the warranty.

(*b*) Provide warranty execution training as an integral part of the fielding/new equipment training process for the item with emphasis on procedural differences that may be required due to geographic, organizational, or mission differences of the using units.

(*c*) Ensure that a cost-benefit analysis is performed to determine the cost effectiveness of a proposed warranty.

(*d*) Document and retain in the Master Acquisition Plan (MAP) and contract file (see reference (c)), the cost benefit analyses performed in the decision process to acquire or not acquire a warranty for the acquisition.

(*e*) Request waivers for warranties on weapon systems/equipment when the proposed warranty is not cost-effective or in the best interest of the Government. Copies of those requests shall be maintained as part of the program's documentation. Such waivers shall include the following information as a minimum:

1. A description of the system and its state of production as well as the number of units delivered and anticipated to be delivered during the life of the program.

2. The specific warranty or warranties for which the waiver is requested, the duration of the waiver (if it extends beyond the contract under consideration), and the rationale for the waiver. Include in the rationale a statement describing the cost effectiveness of the warranty. This statement shall reference the analysis performed and documented to substantiate it.

3. A summary of the assumptions, cost factors, benefits, and conclusions contained in the cost benefit analysis. Identify who performed the analysis.

4. A description of the techniques to be employed to assure acceptable field performance of the weapon system.

(*f*) Ensure that procurement work orders (PWO) contain sufficient information on the equipment and the warranty desired to develop a warranty clause, a copy of the proposed warranty provision, or a copy of the approved waiver of the warranty requirement.

*(*g*) Provide the CG MCRDAC with recommended essential performance characteristics.

*(6) Approve the essential performance characteristics to be warranted by the contractor.

c. The CG MCLB Albany shall:

(1) Establish a warranty information data base for the collection and tracking of warranty claim and usage data to provide the information required for warranty assessments and reporting.

(2) Act as the Marine Corps warranty administrator between the Marine Corps and contractor or administrative contracting officer (ACO)/principal contracting officer when a weapon system and/or components are to be supported by commercial or negotiated warranties.

(3) Ensure that the procedures in enclosure (2) are used to notify contracting officers of warranty claims.

* (4) Review draft ALOs to ensure the adequacy of warranty information. Provide comments and recommendations to the CG MCRDAC (PSI-L) to correct any deficiencies identified.

* (5) Receive, from contractors, monetary reimbursements and parts resulting from warranty claims.

* (6) Conduct trend analyses of QDRs submitted per reference (f) to determine if warranted essential performance characteristics of the weapon system and/or components are being met. Advise the CG MCRDAC (PSI-G and the PM) of those instances where the trend

analyses indicate the specified essential performance characteristics are not being met.

* (7) Forward the results of the trend analyses of QDRs to the CG MCRDAC (PSI-G) along with a determination of whether or not the failures are the result of design or manufacturing defects.

* (8) As the warranty administrator collects the information required in paragraph 4s, preceding.

* (9) Submit a consolidated report of warranty claim and usage data in the format described in enclosure (4) to the CMC (L) and the CG MCRDAC (PSI-L) 15 days after the end of the 2d quarter (for the period 1 January through 30 June) and the 4th quarter (for the period 1 July through 31 December). This report has been assigned Report Control Symbol MC-4105-01.

(10) Program or budget funds to administer the warranty program and for repair/replacement of the support items determined after negotiations with the contractor to be excluded from coverage by the warranty.

**d.* The CGs of the Fleet Marine Forces (FMFs), 4th Marine Division (MarDiv), 4th Marine Aircraft Wing (MAW), and Marine Corps Bases (MCBs) shall:

(1) Ensure procedures are established down to the using unit implementing the warranty claim procedures identified in enclosure (2) of this Order.

(2) Designate a point of contact for an installation of predetermined command/geographical area. The number of personnel/units contacting the contractor or dealership must be kept to a minimum to preclude conflicting resolution of warranty matters. Therefore, a warranty coordinator shall be appointed by the CG FMF as the point of contact within each FMF. The FMF warranty coordinator will ensure that warranty coordinators are appointed at commands possessing fourth echelon maintenance capabilities.

(*a*) Continental United States (CONUS) or outside continental United States (OCONUS). Active U.S. Marine Corps units shall process warranty claims through appropriate support and maintenance channels to the warranty coordinator.

(*b*) SMCR. Reserve units possessing organizational maintenance capability, which are geographically separated from intermediate maintenance activities are authorized to make warranty determination and to coordinate warranty actions with the warranty administrator at MCLB Albany. Reserve units not possessing organizational maintenance capability will obtain warranty service through a supporting organizational maintenance activity.

(3) Ensure warranty claims are filed for all failures of warranted items.

(4) Ensure warranty coordinators are the focal point for coordinating all warranty actions between the using unit and local dealers or manufacturers, the warranty administrator, and contracting officers.

(5) Execute warranty procedures as described in ALOs.

(6) Maintain files and records as necessary to manage the warranty program for weapon systems and locally procured equipment.

(7) Ensure warranty coordinators provide information to the using units on warranty coverage and exclusions, clarify warranty claim issues, and provide assistance to implement the system warranties.

6. *Reserve Applicability.* This Order is applicable to the Marine Corps Reserve.

J. J. WENT
Deputy Chief of Staff
for Installations and Logistics

DISTRIBUTION: E

Copy to: 7000062, 106, 144, 148, 160/8145001

DEPARTMENT OF THE NAVY
HEADQUARTERS UNITED STATES MARINE CORPS
WASHINGTON, D.C. 20380-0001

MCO 4105.2 Ch1
LA-PSI-dt
12 Apr 1988

MARINE CORPS ORDER 4105.2 Ch 1

From: Commandant of the Marine Corps
To: Distribution List

Subj: Marine Corps Warranty Program

Encl: (1) New page inserts to MCO 4105.2

1. *Purpose.* To transmit new page inserts to the basic Order.
2. *Background.* As a result of the changes to the Marine Corps acquisition organization and the activation of the Marine Corps Research, Development, and Acquisition Command (MCRDAC) specific responsibilities in the management and execution of the Marine Corps Warranty Program must be reassigned.
3. *Action*
   a. Remove present pages 5, 6, and 9 through 12 of the basic Order and replace with corresponding pages contained in the enclosure hereto.
   b. Remove present pages 3, 4, 7, and 8 of enclosure (2) and replace with corresponding pages contained in the enclosure hereto.
4. *Change Notation.* Paragraphs denoted by an asterisk (*) symbol contain changes not previously published.
5. *Filing Instructions.* This Change transmittal is filed immediately following the signature page of the basic Order.

J. J. WENT
Deputy Chief of Staff
for Installations and Logistics

DISTRIBUTION: E

Copy to: 7000062, 106, 144, 148, 160/8145001

DEFINITIONS

For the purpose of this Order, the following terms are defined:

1. *Acceptance.* The act of an authorized representative of the Government by which the Government assumes ownership of supplies tendered or approves specific services rendered as partial or complete performance of the contract.
2. *Commercial Warranty.* A warranty offered by a contractor that sells a substantial amount of the product being acquired by the Government to the general public. The warranty price is generally inseparable from the price of the item and there is no tailoring of the warranty provisions at the time of sale. An example of a commercial warranty would be the 90-day parts and labor warranty provided in the purchase of a new television set.
3. *Cost Benefit Analysis.* A process used to compare the total costs of a warranty with the benefits to be derived from that warranty. This analysis shall be conducted to identify the costs for the life cycle of the item both with and without a warranty. (Note: the DoD Life Cycle Cost Model is capable of performing that comparison. The difficult task is to identify all the associated costs and benefits and placing a dollar value on them for comparison purposes.)
4. *Defect.* Any condition or characteristic in any supplies or services furnished by a contractor, under a contract that is not in compliance with the requirements of the contract.
5. *Design and Manufacturing Requirements.* Structural and engineering plans and manufacturing particulars, including precise measurements, tolerances, materials, and finished product tests for the weapon system being produced.
6. *Essential Performance Characteristics.* Operating capabilities and reliability and maintenance characteristics of a weapon system/subsystem/component that are determined by the sponsor to be necessary for the system to fulfill the military requirement for which it was designed. Usually limited to three to seven characteristics that are readily measurable in an operational environment, though the number may be more if the complexity of the equipment warrants.
7. *Failure-Free Warranty.* A failure-free warranty requires a period of failure-free usage. Commercial and trade practices warranties are examples of this concept. (Note: under this concept, each claim is subject to the contract remedy during the warranty term. Since failures may occur, the cost of the warranty will normally include the expense of repair or replacement that can be expected during the warranty term. This cost may be included in the item price and not identifiable as a separate cost. This type of warranty may be more appropriate when an item's reliability is unknown or unspecified as in the case of a nondevelopment item. Administration costs usually increase because the warranty claims are processed at a lower level in the maintenance chain.)
8. *Incentive Warranty.* A type of warranty that provides incentives for the contractor to exceed minimum design, quality, or performance levels. For example, the contract may establish increasingly higher reliability levels

above the minimum requirement with monetary rewards for the contractor should his system meet these higher standards. (Note: depending upon the structure of the warranty, this may or may not meet the requirements of the Defense Procurement Reform Act.)

9. *Master Acquisition Plan.* The principal planning document for each Marine Crops acquisition program. It describes the proposed system, provides a historical summary, provides guidance for each detailed supporting plan and includes a list of program objectives and milestones. Additional information concerning format and contents of this plan may be found in MCO P5000.10B.
10. *Mature Production.* Follow-on production of a weapon system after manufacture of the lessor of the initial production quantity or one-tenth of the projected total production quantity.
11. *Performance Assurance Warranty.* Term used to describe a warranty in which the primary intent is to assure that minimum design, quality, and performance levels are achieved. (Note: the Government is not seeking anything more than what the contract specifies, and the warranty concept and terms and conditions do not provide any incentives for the contractor to do otherwise. This is the type of warranty required by the new Defense Procurement Reform Act described earlier.)
12. *Prime Contractor.* A party that enters into an agreement directly with the United States Government to furnish goods or services.
13. *Transition Plan.* A plan which depicts those significant events and timing of those events to assure the orderly transition of supply support from the contractor to the Marine Corps Supply System.
14. *Warranty.* A promise or affirmation given by a contractor to the Government regarding the nature, usefulness, or condition of the supplies or performance of services furnished under the contract.
15. *Warranty Administrator.* An individual within a weapon system/equipment management (WS/EM) team who has total management responsibility for all warranty claims/actions regarding a specific weapon system/equipment.
16. *Warranty Coordinator.* An individual assigned responsibility for coordinating warranty actions/functions required between the user and the warranty administrator. (Note: a warranty coordinator will be appointed by the CG FMFs and serves as the point of contact within the FMF on warranty issues. Warranty coordinators appointed below the FMF level will normally be located at the force service support group (FSSG), units possessing fourth echelon maintenance capability, or units whose geographic location mandates an independent warranty coordination capability.)
17. *Weapon System.* A system or major subsystem used directly by the Armed Forces to carry out combat missions. (Note: the term includes, but is not limited to the following, if intended for use in combat missions, tracked and wheeled combat vehicles; self-propelled, towed and fixed guns, howitzers, and mortars; helicopters; naval vessels; bomber, fighter, reconnaissance and electronic warfare aircraft; strategic and tactical missiles including launching systems; guided munitions; military surveillance, command, control, and communication systems; military cargo vehicles and aircraft; mines torpedoes; fire control systems; propulsion systems;

electronic warfare systems; and safety and survival systems. This term does not include related support equipment, such as ground handling equipment, training devices and accessories; or ammunition, unless an effective warranty for the system would require inclusion of such items. This term does not include items sold in substantial quantities to the general public as described in the Federal Acquisition Regulation 15.804-3(c)).

STANDARD WARRANTY PROCEDURES

1. *Purpose.* Certain procedures must be followed by the user of equipment under warranty contracts to ensure the warranty claim system agreed upon between the Marine Corps and the contractor will function as intended. These generic procedures are intended to describe the principal features of the warranty provisions of the equipment under warranty, to provide instruction defining the process of securing warranty services and/or parts covered under the warranty, and to illustrate the proper method of processing warranty claims for service and/or parts. Specific warranty procedures tailored to individual equipment will be included in the applicable contract and promulgated in the equipment's ALO.
2. *Guidance.* Maximum cooperation between contractors, or their representatives, and the warranty administrator at MCLB Albany is desired and necessary. The warranty coordinator should not participate in warranty disputes. Warranty disputes should not cause repair of equipment to be held in abeyance pending resolution of disputes. Follow the local standing operating procedures (SOP) and the procedures detailed in this document when there is sufficient evidence that a warranted part is defective and that replacement parts and/or services or reimbursement is due the Marine Corps. All disputes will be transmitted from the warranty coordinator to the warranty administrator at MCLB Albany for evaluation and review. Disputes requiring resolution will then be forwarded to the contracting officer for appropriate action.
3. *General Equipment Warranty*
   a. A weapon system contract requires three specific warranties, one covering design and manufacturing requirements, one covering defects in materials and workmanship, and one covering essential performance requirements delineated in the contract.
   b. A warranty does not cover conditions resulting from misuse, failure to perform scheduled maintenance, or improper preservation during equipment storage. The warranty does not cover the replacement of consumable/expendable items (such as filters and lubricating oils) used in connection with normal maintenance services.
   c. Upon receipt of the equipment, or as appropriate, the commencement dates of the warranty must be recorded in the remarks portion of the equipment record jacket NAVMC 696D (Motor Vehicle and Engineer Equipment Record Folder) or as directed if the equipment record jacket is not used; i.e., if the Weapons Record Book Part 1 is used in lieu of the record jacket.
   d. Prior to placing new equipment in storage and again at the time of its

withdrawal from storage, the contractor must be notified through the warranty administrator at MCLB Albany. For this action, use the equipment storage report formats which are provided with each end item when it leaves the contractor's facility. An equipment storage report must be partially prepared for each newly delivered equipment placed in government storage, and completed when each equipment is removed from storage and placed in service. It must be prepared properly and submitted within the following time schedules so the Government can fully realize the intended warranty benefits:
(1) In storage—15 days
(2) In service—5 days

e. In the event of a warranted failure, the warranty coordinator may be required to deliver the equipment to an authorized dealership or warranty service shop.

4. *Notification of Warranty Defect*

a. The using unit will immediately notify the warranty coordinator when a warranted item has failed. The warranty coordinator at the designated command/area shall notify the warranty administrator immediately thereafter. Such notification may be either telephonic or in writing. Any telephone notification will be followed by an SF 368 (Product Quality Deficiency Report) prepared per the current edition of MCO 4855.10. An information copy of the written notification, SF 368, will be provided to the FMF warranty coordinator. When repair is being accomplished by the Marine Corps, it will be so stated on the SF 368.

b. Warranty coordinators will receive copies of all warranty-related SF 368 message QDRs. They will have the responsibility for the planning, execution, and monitoring of all warranty matters within the designated command/area. They will possess an overall perspective of the warranty related problems of the using units within the designated command/area.

c. The warranty administrator shall notify the contracting officer within 5 working days after notification of a defect.

d. The contracting officer shall provide disposition instructions to the warranty administrator within 5 working days after receiving the notification of defect.

e. Upon receipt of disposition instructions from the contracting officer or contractor, the warranty administrator will notify the appropriate command of required actions within 2 days.

f. Under the warranty, the Marine Corps should normally have the unilateral right to effect its own repair. If the Marine Corps elects to effect warranty repair or replacement itself, the following will be done:

(1) The warranty coordinator will notify the warranty administrator within 10 working days after repair is complete. This notification will include the original SF 368 and the pink/photo copy of the equipment repair order (ERO)/ERO shopping list (EROSL) associated with that ERO.

(2) The contracting officer or the contractor shall be notified by the warranty administrator within 30 days after discovery of the defects per paragraphs 4a and b, preceding.

(3) The contracting officer or contractor shall provide disposition instructions to the warranty administrator within 5 days after receiving initial Marine Corps notification of defect. The warranty administrator shall take action as appropriate.

(4) When parts replacement is required, the contractor shall respond within 5 days of its intention to furnish identified parts and shall provide same within 5 days after receipt of notice by the contracting officer/warranty administrator.

5. *Storage Procedures.* Specific tasks to be performed before planing an item in storage and while the item is in storage shall also be identified in the ALO.

6. *Safety Recall*

a. If a safety recall occurs during the equipment warranty period, the contractor shall, per the contract, extend the term of the warranty for each piece of equipment on an item-by-item basis, by a period equal to the time required to make necessary safety defect corrections on each piece of equipment. Extensions of warranty coverage shall be annotated in the remarks section of the equipment record jacket or as directed if equipment record jackets are not used.

**b.* Once it has been determined by the contractor that a problem is safety related, it shall be the responsibility of the contractor, as defined by the terms of the contract, to furnish a defect information report to the CMC (L), CG MCRDAC, and MCLB Albany (WS/EM), for each defect in the equipment produced under the applicable contract. This report shall be submitted within 5 working days after the defect on the equipment or components have been identified.

c. It shall be the responsibility of the contractor, as defined by the terms of the contract, to maintain a record of equipments initially shipped to consignees identified on the DD Form 250 (Material Inspection and Receiving Report).

d. The contractor, as defined by the terms of the contract, shall remedy safety defects or failures, including the replacement or correction of defective parts in the Government inventory, and shall provide the Marine Corps with any reports required during the remedy process.

e. Additionally, the contractor, as defined by the terms of the contract, shall provide all the necessary instructions for the Government to implement the remedy process, including the information required for the Marine Corps to determine the impact of the remedy process on its publications. The information regarding the remedy process will be in a format similar to that of modification instructions (MIs) or technical instructions (TIs).

7. *Warranty Dispute Claim*

a. Definition. Failure of the Marine Corps and a contractor to agree on who is responsible to repair/replace any item submitted per the warranty procedures shall be a dispute concerning a question of fact within the meaning of the disputes clause of the contract.

b. Dispute Settlement. In situations where the contractor declines to repair or replace items for which the Marine Corps believes itself to have a valid warranty claim, or when the contractor furnishes parts and services to the Marine Corps and later claims that replaced parts were not damaged due to defect in design, materials, and workmanship; a settlement will be reached through the contracting officer as follows:

(1) Contractor declines repair.

(*a*) When a contractor, or an authorized dealer declines to repair an item under warranty, the user should notify the warranty coordinator and proceed to repair the item. Normal supply and maintenance procedures should be used.

(*b*) The warranty coordinator shall immediately report the situation by message to the MCLB (Code 856) Albany with an information copy to the user, per MCO 4855.10, as follows:

1. Identify equipment and reference original SF-368 reporting defect.

2. Record "warranty dispute" and a complete description of the failure.

3. Enter name, activity, and telephone number of person submitting the warranty dispute.

4. Enter name, address, and telephone number of the contractor representative or dealership that refused the service.

5. Enter specific reason(s) given for refusal.

6. Enter the specific facts/ evidence that will refute the contractor's reason(s) for refusal, including photographs and sketches, if possible.

(*c*) The warranty administrator shall forward warranty disputes submitted by the warranty coordinator to the contracting officer for resolution with the contractor.

(2) Contractor requests reimbursement.

(*a*) When the contractor makes an analysis, and claims that part(s) failure was not due to defective workmanship, materiel, or design deficiency; the Government will be invoiced for all costs and expenses incurred.

(*b*) If the contracting officer decides the contractor's claim is valid, the warranty administrator will be notified.

8. *Cash Reimbursement From Contractors*

a. Cash reimbursement from contractors shall be considered an overpayment on a public voucher and shall be collected per paragraph 043108 of the Navy Comptrollers Manual (NavCompt Manual).

b. Any proceeds resulting from a reduction in the contract price as represented by a cash refund will revert to the appropriation concerned.

c. When collecting the reimbursements the DD Form 1131 (Cash Collection Voucher) will be prepared per paragraph 047223-2 of the Nav Compt Manual. To properly prepare the voucher the warranty administrator must ensure that the appropriation data associated with the warranted system is included on the voucher. The APO will provide that information to the warranty administrator in the ALO. In addition to the copies of the voucher necessary to process it through the disbursing channels the warranty administrator shall ensure that a copy is forwarded to the APO for the system to ensure the reimbursement is credited to the proper account. The warranty administrator shall retain a copy of the voucher as part of the information base to be provided in the Warranty Usage Report.

9. *Government Forms*
   a. Record the commencement date of the warranty in the appropriate equipment record (refer to the ALO for appropriate form to be used).
   b. SF 368. Prepare a message in SF 368 format and forward to the MCLB (Code 856) Albany per MCO 4855.10, with an information copy to the warranty coordinator.
   c. NAVMC 10925 (EROSL). Use this form as a source document to report repair parts used/provided by the warranty dealership, using a "WP" advice code in order to establish demand/usage history. Ensure this usage data is reported to Marine Corps Integrated Maintenance Management System/Supported Activities Supply System (MIMMS/SASSY) per the current edition of UM 4790-5 and the following:
      (1) The purpose of the advice code "WP" is to administratively record usage data on warranty parts requisitioned "off-line" from nonsystem sources.
      (2) When consumable repair parts for a warranty item are required, the using unit shall submit a MIMMS "4" parts transaction with a "WP" advice code. This transaction will generate usage data via a "DHA" but will not pass a requisition to SASSY. The actual requisitioning of the required parts will be accomplished per the instructions provided by the warranty administrator.
      (3) Upon receipt of the warranty part, the using unit shall submit a MIMMS "8" parts transaction with the authority code of "2" to indicate "receipt" for the item on the Daily Process Report and close the parts trailer while capturing lead time data.
      (4) For secondary reparables, the maintenance facility effecting repair shall submit the appropriate "4" and "8" part transactions using the secondary reparable national stock number.
   d. *NAVMC 1018 (Inspection/Repair Tag).* Use this form to tag defective parts to be returned to the contractor, per TM-4700-15/1. Include the SF 368 number on the tag.
   e. *NAVMC 10245 (Equipment Repair Order (ERO)).* Prepare per TM-4700-15/1 and provide the pink/photo copy with the returned parts. Ensure usage data is reported in the MIMMS AIS per UM 4790-5.
   f. *Equipment Storage Report.* The contractor shall provide the blank report formats as shown in Figure 1 to the Government representative prior to equipment removal from plant. The forms shall be completed as follows (for each equipment shipped and distributed):
      (1) Part I is completed by the Government representative when the end item leaves the contractor for the storage facility.
      (2) Part II is completed by the unit representative when the equipment is placed in storage. One copy of part II will be provided to/for:
         (*a*) The contractor (Attn: Warranty Administrator).
         (*b*) The ACO/PCO.
         (*c*) The equipment.
         (*d*) The unit files.

(*e*) CG MCLB Albany, GA 31704-5000 (Attn: Warranty Administrator, Code WS/EM).

(3) Part III is completed by the unit representative when the equipment is removed from storage. One copy each to:
   (*a*) The contractor (Attn: Warranty Administrator).
   (*b*) The ACO/PCO.
   (*c*) The equipment.
   (*d*) CG MCLB Albany, GA 31704-5000 (Attn: Warranty Administrator, Code WS/EM).
   (*e*) The CMC (LM).
   (*f*) The unit files.

I. Equipment data

A. Contract number________________

B. Equipment serial number________________

C. DD 250 acceptance date ________________

D. DD 250 shipment number ________________

E. Manufacturer's serial number________________

F. Type of storage program: MO___ CRSP___ DEPOT___ MPS ___

II. Depot storage entry data

A. Location ________________

B. NSN________________

C. Storage date________________

D. Equipment mileage ________________

E. Date report forwarded to contractor________________

F. Depot representative signature________________

G. Type of storage program: MO___ CRSP___ DEPOT___ MPS ___

III. Depot storage removal data

A. Removal data________________

B. Equipment mileage ________________

C. Final destination________________

D. Date report forwarded to contractor________________

E. Depot representative signature________________

Figure 1 Equipment Storage Data

EXPECTED FAILURE CONCEPT

1. The expected failure concept is based upon the premise that the Marine Corps acquires weapon systems to satisfy a stated requirement. Specifically, the Marine Corps will identify a minimum level of reliability for the system being acquired. This reliability will usually be expressed in terms, such as mean time between failure or failure rate and operating hours for the system. During the design of the system, the developing contractor will allocate reliability requirements to subsystems, components, and piece parts that make up the system. Because of limitations (which include cost, technology, and materials) that exist in the acquisition of a system the Marine Corps seldom, if ever, requires a system to possess 100 percent reliability. The Marine Corps recognizes and plans for periodic equipment failures; however, the Marine Corps wants to ensure that these failures do not exceed those normally expected when a certain level of reliability is specified and the system is being utilized in the operating cycle designed for. As long as the system does not exceed the number of failures expected, the contractor has met the specified reliability and should not be held liable if the system fails. When the failures exceed the number expected, the contractor has failed to meet the requirements of the contract and the Marine Corps should seek corrective action for the deficiency.
2. To apply this concept to warranty requirements for the system, first, determine the desired duration of the warranty. Be sure to include in that determination any time that system will be in storage after acceptance by the Government. Next, determine the operating hours during the warranty period. Using that figure and the reliability value specified in the contract, calculate the expected number of failures for that system during the warranty period. Multiply that figure by the number of systems covered in that production lot. The result is the expected number of failures for the system under warranty. This figure is the threshold that must be breached to invoke the warranty. As long as the number of failures is below the threshold, the contractor is not liable; when the threshold is breached each failure becomes a warranty claim. Expected failure thresholds should be determined for all components reparable at fourth echelon or higher to be covered by the warranty. The warranty coordinator will report all failures of warranted items to the warranty administrator, who will track the failures and determine when the threshold is breached. Once breached the warranty administrator invokes the warranty by notifying the contracting officer using procedures detailed in enclosure (2). An example follows:

   SU = systems usage (hours, miles, etc.)
   MTFB = mean time between failure
   # SYS = number of systems in production lot
   # F = number failures

   SU = 20,000
   MTFB = 1000 h
   # SYS = 100
   # F = x

   SU/MTBF × # SYS = x

$20{,}000 \div (1000 \times 100) = x$ or 2000 is the expected number of failures for the system. When the 2001st failure is recorded then the warranty administrator would start submitting warranty claims.

Warranty Usage Report
Nomenclature, NSN, Model #
Report Symbol MC-4105-01

Contract number (include lot #)	Contractor name	Federal supply code of manufacturing (FSCM)	ALO no. and date	Serial lot registration # (range)	Warranty duration	Usage limits (hrs/rds/ miles)

Start date of first item Warranty Period	End date of last item Warranty Expiration	Contract cost of warranty and item cost	Claim Data				
			# Submitted	$ Honored	$ Disputed	$ Denied	Reason for denial

Note: If desired attached a separate remarks page.

Appendix

H

Government Accounting Office Report on DOD Warranties

Executive Summary

Purpose

The Congress passed warranty laws in 1983 and 1984 because of its concern that weapon systems often failed to meet their military missions, were operationally unreliable, had defective and shoddy workmanship, and could endanger the lives of U.S. troops. These laws require defense contractors to guarantee that weapon systems will meet performance requirements specifically delineated in the contract. It is the Department of Defense's (DOD) policy only to obtain warranties that are cost-effective. Because the services spend hundreds of millions of dollars on warranties each year, GAO reviewed DOD's warranty program to determine whether the services (1) had effective warranty administration systems and (2) were performing cost-effectiveness analyses as required by DOD and service regulations.

Background

The Secretary of Defense has delegated administration of the warranty program to the military services. The services are responsible for issuing implementing rules, regulations, and procedures pertaining to warranties. Procurement activities within the services are each responsible for warranty design and administration activities.

The current law requires warranties on weapon systems that have a unit cost of more than $100,000 or an expected total procurement cost of more than $10 million. However, the Secretary of Defense may waive this requirement if it can be shown that the warranty is not likely to be cost-effective. Both DOD and service regulations require cost-effectiveness analyses of proposed warranties. In addition to cost-effectiveness analyses, Army and Air Force regulations require an assessment of warranties while they are still active and a post-warranty analysis to measure the results actually achieved. A Navy instruction requires annual collection and analysis of actual warranty use and claim information.

Results in Brief

The Office of the Secretary of Defense (OSD) is not actively overseeing warranty administration by the services. And, the services have not yet established fully effective warranty administration systems. As a result, DOD has little assurance that warranty benefits are being fully realized.

Waivers of warranty law requirements generally are not being sought by the procurement activities included in GAO's review. Problems are being experienced in performing cost-effectiveness analyses, thus, the activities are not in a position to know whether they should seek waivers.

Principal Findings

OSD does not actively oversee warranty administration. OSD is not actively overseeing the services' progress in establishing effective warranty administration systems. The focal point for warranty administration has been delegated to each service, and OSD functions only in a reactive mode to deal with issues raised by audit groups and other interested organizations.

Fully effective administration systems have not been established. The services are in various phases of establishing systems to administer their warranties. The Navy issued a policy on the use of warranties in 1987, but has not defined roles and responsibilities or established overall procedures and controls for administering warranties. The Air Force recently issued comprehensive guidance and is in the process of establishing its system. The Army has issued policies and procedures, but the Army procurement command visited by GAO is experiencing problems in executing them.

Adequate cost-effectiveness analyses are not being prepared. Procurement activities included in GAO's review either have not been performing cost-effectiveness analyses or have prepared analyses that do not adequately support conclusions that proposed warranties are cost-effective. As a result, procurement activities were not considering waiver requests in their decisions on proposed warranties because their analyses did not provide a convincing basis to support requests for waivers in cases where warranties may not be justified because they would not be cost-effective.

Post-warranty evaluations are not being prepared. To achieve full benefits from weapon system warranties, DOD needs assurance that the warranties are accomplishing their purpose. A system that provides information to evaluate actual warranty benefits is a key element in effective warranty administration and could provide such assurance. The procurement activities GAO visited have not yet evaluated warranty benefits after warranties have expired. The general lack of evaluations has been due to problems experienced in establishing effective warranty administration systems that will provide the information needed to perform post-warranty evaluations. These problems included delays in establishing warranty information collection procedures and difficulties in obtaining accurate information concerning warranty claims.

Recommendations

GAO recommends that the Secretary of Defense expand his oversight role in warranty administration by establishing milestones for the services to meet in implementing warranty systems and ensuring that the services consider the use of waivers as viable options when it can be shown that a warranty is not cost-effective.

Agency comments

DOD did not agree that the Secretary of Defense should establish milestones for the services to meet in implementing warranty administration systems, but stated that it would request status reports from the services on their efforts to implement administration system (see App. A). While status reports will be helpful, GAO continues to believe that because the services' progress in establishing effective warranty administration systems has been slow, OSD needs to identify milestones for the completion of the generally accepted elements of a warranty administration program tailored to the status of each service's program.

DOD agreed with GAO's recommendation to emphasize the use of waivers when it can be shown that a warranty is not cost-effective.

Appendix

I

Warranty Checklists

Warranty Development Checklist

Program risks and goals

- Has relevant documentation been consulted to identify the significant program risks and requirements in order to develop a meaningful warranty?

Warranty requirement

- Have DOD and service policies been used to verify that a weapon system warranty is required?
- If the requirement for a weapon system warranty is not established, would it still be a good idea to have a warranty to meet program goals or diminish program risks?

Warranty coverage

- Have warranties for defects in materials and workmanship and for conformance to design and manufacturing requirements been included?
- Have candidate EPRs such as reliability, maintainability, and operation performance parameters been considered?

If there is an essential performance requirement, is it

- Consistent with the specification?
- Not easily measured in the laboratory on a one-time basis, but should be covered by a warranty in the field?

- Measurable in the field without dispute?
- Translatable to a meaningful remedy in case of failure to comply?
- Controllable to a reasonable extent by the contractor?

Warranty strategy

- Has a warranty strategy been devised that considers such aspects as competition, contractor bid of guarantee values, contractor comment on draft warranty provisions, and warranty RFP language and proposal evaluation?

Warranty scope

- Does the warranty clearly identify what units are included, and what units are excluded, if any?

Duration

- Has a realistic and reasonable duration for the warranty been determined?
- If the warranty ends at different time for each item, will this cause implementation problems?
- If the warranty duration is related to population hours, such as total flying hours, can accurate measurement be made?

Presumption of coverage

- Is the "presumption of coverage" language used to minimize potential disputes?

Exclusions

- Are there reasonable exclusions from warranty coverage, such as acts of God and combat damage, in order to protect contractor from undue risks?

Contractor repair

If the warranty requires contractor repair:

- Is contractor repair acceptable in view of current service capability and mission criticality?
- Can the warranty units be easily shipped within the terms of the warranty?

- Can unauthorized maintenance be controlled?
- Can good returns be minimized?
- Is there a control on contractor repair turnaround time?
- Are there reasonable data requirements placed on the contractor to provide repair and failure analysis data?
- Have plans been made to monitor contractor warranty repair performance?

Field measurement

If the warranty requires field measurement to verify conformance to an EPR:

- Can existing service data collection systems be used?
- If there are no adequate existing data systems, have plans been made and approved to implement a new system?
- Is using field measurement data to determine conformance to an EPR better than using a special verification test?
- Has a data collection and analysis plan been developed that clearly defines responsibilities, collection periods, and analysis procedures?

Dormant systems

If a warranty is related to dormant system performance:

- Are there long nonuse periods during which deterioration is possible?
- Are there enough periodic tests performed to measure storage performance?
- Are there provisions to allow the contractor to monitor storage performance tests?

Cost-benefit analysis

- Have cost-benefit analyses been performed on a timely basis?
- Do the results of the cost-benefit analyses adequately support the warranty decisions that have been made?

Remedies

- Have remedies been developed that are equitably related to the degree of warranty breach?

- Should there be a limit on the total contractor liability such as a cost ceiling related to the total contract value?
- Is redesign a specific remedy?
- If reimbursement for government repair is a remedy, are there specific means to determine the amount of the contractor's liability?

Marking

- Has marking of warranted items been specified to ensure proper handling and disposition in the field?

Technical data

- Do requirements for technical data include adequate reference to the warranty?

Training

- Does planned training include coverage to implement and manage the warranty?

Transportation

- Does the warranty state who is responsible for transportation costs?

Implementation

- Have all possibilities been considered so that support units can unambiguously determine if a warranty breach has occurred?
- Does the warranty ensure that unacceptable burdens will not be imposed on the user and support communities?
- Has a warranty administration or implementation plan been developed?

Warranty Administration Plan Checklist

Introductory material

- Have the effective date and duration been identified?
- Has coordination been completed?

Acquisition background

- Is the purpose of the acquisition program clearly stated?

- Is the purpose of the warranty clearly stated?
- Is a brief history of the acquisition program included?
- Is a brief rationale of the warranty selection included?
    - Are the cost-benefit considerations clearly explained?
    - Is the EPR selection rationale clearly explained?
    - Is the rationale for the remedies included?

Weapon system warranty terms

- Are complete warranty terms included?
    - What is warranted?
    - How are warranted items identified?
      –Include illustration of markings if possible.
      –Reference technical publications if applicable.
    - How long does the warranty last?
    - What are the remedies?
    - How will data be gathered, recorded, and exchanged?
    - What are the contractor obligations?
    - What are the government obligations?
    - What are the exclusions?

Cost-benefit analysis

Are the following referenced?

- Cost-benefit analysis
- Methodology
- Data
- Effectivity

Are the following facts about the analysis included?

- Limitations
- Assumptions
- Data accuracy

Are the conclusions clearly supported?

Is an update of relevant events since the last cost analysis included?

Warranty administration

- Are the warranty responsibilities of each organization, including the contractor, separately listed?
  - Are the responsibilities included in other documents also listed, for example, deficiency reporting?
  - Are the controlling documents referenced such as public law, regulations, memorandums of agreement, and the contract?
- Are the information flow paths clearly defined? (A flow diagram may be useful.)
- Are reasonable suspense times levied?
- Is hardware disposition clearly defined?

When applicable, does the plan address:

- Post-warranty-period activities, such as configuration updates, transition to organic maintenance, and assessment of warranty benefits?
- On-equipment (organizational-level) maintenance procedures? (Cite only exceptions to standard procedures.)
- Off-equipment maintenance procedures (for intermediate, direct support, and general support levels)? (Cite only exceptions to standard procedures.)
- Depot maintenance procedures? (Cite only exceptions to standard procedures.)
- RTOK processing?
- Maintenance data requirements? (Cite only exceptions to standard procedures.)
- Other maintenance exceptions such as FMS and special-use assets?
- Transportation procedures? (Cite only exceptions to standard procedures.)
- Contractor data and reporting requirements?
- Special packaging requirements?
- Damage reporting?
- Special storage requirements (resulting from warranty only)?
- Commingling of warranted and unwarranted assets?
- Operation of contractor secure storage area?
- Consideration of stock-issue priorities?

- Communications procedures for maintenance and utilization data? (Cite only exceptions to standard procedures.)
- Description of required contractor in-plant procedures?
- Custody-transfer requirements?
- ECP processing procedures? (Cite only exceptions to standard procedures.)
- Configuration control procedures? (Cite only exception to standard procedures.)
- Warranty impacts on technical orders?
- Warranty funding?
- Funding for repair of exclusions?

Warranty team

Is the warranty team defined?

Is complete identification given, including

- Complete title?
- Brief description of team duties?
- Telephone numbers—defense switched network (DSN) and commercial?
- Addresses?

Program management responsibility

- Is the warranty administration plan consistent with any program management transfer responsibilities?
- Are any changes in responsibilities delineated for each organization?
- Are due dates established in relation to any transfer milestones?
- Is a meeting planned as part of any transfer process to discuss and clarify responsibility changes and procedures?
- Will the contractual warranty provisions, such as CDRL deliveries, require updating as part of any transfer of responsibilities?
- Are updates of memorandums of agreement provided for?

Foreign military sales

- Does the warranty cover any FMS?
- Are there unique FMS warranty considerations?

- Is another complete plan advisable?
- Is it referenced here?

Contractor logistics support and interim contractor support

- Is there any contractor logistics support or interim contractor support?
  - Are contractor warranty responsibilities described?
  - Are contractor warranty responsibilities required by the statement of work?
  - Has the administrative contracting officer (ACO) been tasked to monitor the contractor logistics support/interim contractor support contractor's warranty responsibilities?

    –Are the ACO tasks in Chap. 4 of the warranty administration plan?
  - Does the warranty last longer than the interim contractor support?

    –Is there a transition plan?
    –Are contractor responsibilities during and after interim contractor support clearly stated?
- Are there procedures for a case in which potential conflicts of interest are resolved, for example, if the same contractor is responsible for invoking the warranty and suffering the remedies?
- Are procedures in place to ensure that warranted items are not repaired under interim contract support funding?

Schedule

- Are the major program milestones included?
- Are the warranty milestones that relate to the program milestones included?
  - Warranty beginning and end dates.
  - Special warranty tests.
  - Contract options.
- Are any transfer responsibility milestones included—including warranty transition milestones?

Training

- Is the overall program training plan referenced?
- Is warranty training included with other training where possible?

- Does the training include all individuals who must make warranty decisions?
- Does the training include all individuals and their supervisors whose actions could void the warranty?
- Does training include recognition of warranty markings and their implications?

Index

Administration cost, 111, 117–118, 123
Air Force Regulation (AFR) 70–11, 275–300
Application criteria, RIW (*see* Reliability improvement warranty)
Approaches to warranty costing (*see* Warranty costing process)
Army Regulation (AR) 700-139, 241–270
Asset tracking, 168–172
  flow, 168–171
  identification of assets to track, 171–172
  supplier and customer obligations, 171
Assurance validation function, 13
Assurance warranty, 14
  versus incentive warranties, 14–16
Availability, 39, 69–70, 74–78, 143, 146, 151–153, 157–159
  inherent, 75
  operational, 75–78

Baseline analysis, 96
Beta distribution, 115–116
Bottom-up analysis, 109–110, 114, 128
Bounding, 115
Built-in test (BIT), 20, 80

Commercial warranties, 5–6, 16–18, 36–37, 41–44, 45–47
Consumer warranties, 5–8, 16–18, 36–37, 41–44, 45–47
  defined, 5–6
  versus commercial, 6
  versus government, 6–8
Combination free-replacement and pro-rata warranty, 17, 93
Comparison of assurance and incentive warranties (*see* Assurance warranty)
Concept exploration, definition phase, 33, 188–189
Contractor:
  benefits (RIW), 61–62
  development activities, 200–205
  interpretation strategy, 53–54
    communication, 54
    documentation, 54
    education, 54
    guidance, 53–54
  manager's checklist, 204–205
    planning, 202–204
    team, 200–202
  risks (RIW), 63–65
Cost, 45, 53
Cost-benefit analysis (CBA), 144
Cost components, 110–114
  administration, 111
  depot establishment, 110
  field service, 111–112
  repair, 111
  risk, EPR penalties, 113–114
    redesign and retrofit, 112–113
    repair, 112
  transportation, 111
Cost effectiveness (CE), 5, 133, 143–144, 152–154, 158, 205–207
Cost-effectiveness analysis (CEA) concept, 146–154
  application of LCC, 147
  benefits, 149
  data, 148
  framework, cost effectiveness, 152–154
    life-cycle cost, 150–151
    other cost elements, 154
    system effectiveness, 151–152
  performance of cost-effectiveness analysis, 148–149
  purpose, 146
  team, 147
  timing, 146–147

Cost-effectiveness analysis (CEA) process, 154–162
  analysis performance, 160–161
  data estimation, 158–159
  model formulation, 156–158
  team formation, 154–155
  value decision, 161–162
Cost-effectiveness requirements, 144–146
  DFARS subsection 246.770-7 and 246.770-8, 144–145
  GAO report no. 89-57, 145–146
Cost-risk analysis, 116–127
Cost tracking, 181–182, 184
Coverages, 21, 44
  individual item, 21
  systemic, 21
Critical costing considerations, 107–110
  avoidances, 108–109
  credibility, 108
  cost-risk analysis, 110
  negotiation, 109
  rationale, 109–110
  risk, 109
  thoroughness, 108
  uniqueness, 107–108
Customer satisfaction, 8, 210

Data, 45
Data base, 54
Database management, 171–175
  computer hardware and software, 172
  examples of databases, 175
  inputs and outputs, 172
  procedures, 173–174
Defect-free warranty, 19
Defects, 16
Defense Federal Acquisition Regulation Supplement (DFARS) Subpart 246.770, 233–239
Definition and scope, RIW, 58–59 (*see also* Reliability improvement warranty)
Definitions, 3, 44
  guarantee, 3
  warranty, 3
Demonstration–validation phase, 33, 189–190
Design and manufacturing requirements, 39, 42, 44
Disclaimers, 44
Duration (period of warranty), 44

Engineering–manufacturing development phase, 33, 190–191
Essential performance requirements (parameters), 39–42, 44, 50, 69–79, 112–114, 119–120, 137–141, 193–196, 200
Examples of performance guarantees (*see* Performance guarantees)
Exclusions, 44–45
Expected failure warranty, 18–19
Exponential distribution, 27–29
Express warranty, 35–36

Failure, 16
Failure-free warranty, 18
Failure (hazard) rate, 27–32
Fee (contractor), 127
Field service cost, 111–112, 117, 119, 123
Figure of merit, 95, 143
Fleet warranty, 18, 93
Free-replacement warranty, 16, 92
Functions of warranties, 13
  assurance validation, 13
  incentivization, 13
  insurance, 13

Government Accounting Office (GAO) Report on DOD Warranties, 323–325
Government benefits, RIW (*see* Reliability improvement warranty)
Government development activities, 185–200
  alternatives, 193–197
  concept exploration–definition phase activities, 188–189
  demonstration–validation phase activities, 189–190
  EMD phase activities, 190–191
  plan, 197–199
  preparation of clause, 199–200
  program objectives, 185–186
  selection factors, 191–193
  system life cycle, 186–188
Government risks, RIW (*see* Reliability improvement warranty)
Government warranties, 6–8
Guarantee (guaranty), 3

Iacocca, Lee, 9, 183
Implementation cost, 110–112
Implementation of warranty, 107
Implied warranty, 35–36
Incentive warranty, 14

Incentives, 20–23, 40, 71, 74, 83–86, 112–114, 117, 119
Incentivization function, 13
Individual item coverage, 21
Insurance function, 13

Joint Army–Industry Warranty Working Group (JAIWWG), 205–207
  actions, 206
  Army-industry interaction, 206–207
  charter, 205
  findings, 206
  implementation challenges, 207
  membership, 206
  mission, 205
  structure and process, 205–206
Juran, Joseph, 211

Karrass, Chester, 137

Legislation, warranty, 36–41
  Magnuson-Moss Federal Trade Commission Improvement Act, 36–37
  major weapon system contractor guarantees, 37–41
  uniform commercial code, 36–37
Lessons learned, administration, 182–184
  continuous negotiation, 183
  customer-supplier relationships, 182–183
  importance of cost tracking, 184
  importance of review boards, 183–184
Liability, contractor, 11, 13
Life cycle of system, 33, 186–187
Life-cycle cost (LCC), 97–98, 143, 146–147, 150–151, 156–157, 160–161, 163–164
Limited warranty, 37
Logistics, 75–77, 88, 100–103, 115, 144, 155, 159, 188–191, 199
Logistics support cost (LSC), 20, 72–74, 79–80
Loss (*see* Profit/loss)

Magnuson-Moss Federal Trade Commission Improvement Act, 5–6, 36–37, 215–228
Maintainability, 20, 39, 69–70, 73, 75, 143, 151, 157–159
Maintenance cost, 73–74
Major weapon systems, contractor guarantees, 6, 37–41, 215–217
Management reports, 172–173
Markings, 45, 180
Material and workmanship requirements, 39, 42, 44
Mean administrative delay time (MADT), 75–78, 151, 158–159, 164
Mean downtime (MDT), 75–78, 151, 158–159, 164
Mean logistics delay time (MLDT), 75–78, 151, 158–159, 164
Mean time between failures (MTBF), 20, 23–24, 26–29, 39–40, 57, 61–62, 64, 67–68, 71–72, 73–78, 80–86, 97–98, 100–103, 110, 112–113, 117–118, 120, 136–137, 152, 160–161, 163–165
Mean time between maintenance (MTBM), 77–78, 151, 157, 159–160, 164
Mean time between removals (MTBR), 20, 24, 39, 117, 119, 123–124
Mean time to failure (MTTF), 30–32
Mean time to repair (MTTR), 20, 73–75, 78, 151, 164
Measure of a good warranty, 5
Merchandising, 8
Military warranties, 7, 18–21, 37–41, 44–45, 47–53
Models, formulation of, 72–74, 88, 95–98, 115–123, 149–154, 156–161
Monte Carlo simulation, 110, 116, 120–126

Negotiation, 106–110, 183
  example, 138–142
    negotiation, 140–142
    situation, 138–140
  process, 133–136
    acceptance of quote, 135
    agreement on requirements, 135–136
    agreement on supporting data, 136
    definition of requirements, 135–136
    interpretation of requirements, 134–135
    quotation, 135
    should costing, 135

Operating and support (O&S), cost, 58, 60, 66–67, 147, 150, 156–157
Operation and support phase, 33–34
Operational capability parameters, 20–21, 39–40, 70, 151
Other rights and remedies, 45, 51–52

Penalties remedy, 20–23, 40, 71, 74, 83–86, 112–114, 117, 119
Penalty cost, 112–113, 117, 119, 123
Performance, 2, 209–212
Performance guarantees, 20–21, 79–86
  examples of, 79–81
    built-in test, 80
    logistics support cost, 79–80
    mean time between failure, 80
    shop visit rate, 80–81
    system mission readiness, 80
    turnaround time, 80
Planning, contractor (*see* Contractor)
Potential benefits, RIW (*see* Reliability improvement warranty)
Price, contractor's, 105–106, 127
Price adjustment remedy, 22
Procedure to obtain warranty remedy, 43–44
Production–deployment phase, 33–34
Profit/loss, 8–10
Pro rata warranty, 17, 92

Quality, 2, 209–212
Quality Function Deployment (QFD), 209–210
Quotation process, 114–127
  costing process, 115–127
  establishment of price, 127
  team formation, 115
Quote, 105–106, 127, 133–136

Realities of warranties, 11–13
Redesign remedy, 22, 112–113, 117, 119, 123
Redesign-retrofit cost, 112, 117, 119, 123
Reliability, 13, 20, 69–70, 73, 75, 143, 146, 151–153, 157–159
  considerations, 26–32
    exponential distribution, 27–29
    weibull distribution, 29–32
Reliability improvement warranty (RIW), 57–68
  application criteria, 66–67
  definition and scope, 58–59
  potential benefits, 59–62
    contractor, 61–62
    government, 59–60
  problem areas, 65–66
  risks, 62–65
    contractor, 63–65
    government, 62–63
Remedies, 21–23
  penalties, 22–23
  price adjustment, 22
  redesign and retrofit, 22
  repair and replacement, 21–22
Repair remedy, 21–22, 57–68, 90–91, 111–112, 117–118, 120
Repair considerations, 175–180
  demand on supplier resources, 175–178
  impact on production, 178–180
Repair cost, 111, 117–118, 123
Replacement remedy, 21–22, 57–68, 90–91, 111–112, 117–118, 120
Reporting, 180–182
  cost reports, 181–182
  review board, 182
  status reports, 180–181
Requirements, 36–41, 44–45
Retrofit remedy, 22, 112–113, 117, 119, 123
Risk, 13, 59, 61–65, 106–107, 109–110, 112–116, 120–122, 127, 210–211
  cost, 112–114
  issues, 23–26
    general, 23–24
    particular, 24–26

Scope of warranty, 49
Sensitivity analysis, 89, 95–98, 110, 115, 149, 156, 159, 161, 164–165
Service strategy, 89, 94
Shipping cost, 111, 117–118, 123
Shop visit rate (SVR), 80–81
Simulation model, 91, 110, 116, 120–126
Staffing, warranty administration, 167–168
Structure of requirement, 44–45
  cost, 45
  coverage, 44
  data, 45
  definitions, 44
  duration, 44
  exclusions, 44–45
  markings, 45
  procedure, 44
  remedies, 44
  requirements, 44–45
  other rights and remedies, 45
  transportation, 45
Supportability, 20, 39, 69–70, 73, 75–78, 143, 151, 157–159

System effectiveness (SE), 149, 151–152, 157–158, 164–165
Systemic warranty coverage, 19, 21, 44
System life cycle, 32–34
  concept exploration–definition phase, 33
  demonstration–validation phase, 33
  engineering–manufacturing development (EMD) phase, 33
  production–deployment phase, 33–34
  timing, 34
System mission readiness (SMR), 20, 80

Team, contractor, 200–202
Title 10, Section 2403 of U.S. Code, 6, 37–41, 215–217
Threshold warranty, 18–19, 97, 139–140, 165
Total quality management (TQM), 209–210
Tradeoff issues, 89–95
  administration, 95
  breadth, 93
  competition, 95
  length, 90–91
  service strategy, 94
  type of policy, 92–93
  type of remedy, 93–94
Tradeoff process, 95–98
  define purpose, 95
  establish basis of comparison, 95–96
  identify candidates, 95
  perform baseline analysis, 96
  perform sensitivity analysis, 96–97
  rank candidates, 96–97
  secure data, 96
  select best candidate, 97–98
  select model, 96
Tradeoffs, 62, 65–66, 87–103
Transportation, 45, 51
Transportation cost, 111, 117–118, 123
Turnaround time (TAT), 20, 23, 25, 39, 76, 80–86, 113, 119–120, 123–124
Types of performance guarantees, 69–78 (*See also* Performance guarantees)
  availability, 74–78
  logistics support cost, 72–74
  mean time between failures, 71–72
  operational parameters, 70
Types of warranty policies (*see* Warranty policies)

Uniform commercial code, 5–6, 36–37
United States Code (USC), Section 2403 of Title 10, 215–217

Value of warranty, 8, 24, 41, 136, 143–144, 210
Voice of the customer, 209–210

Waivers, 40, 145–146, 149, 162–163, 165
Warranty, 3, 11–12, 32–34, 92–99, 105–106, 115–123, 127
  cost, 11–12, 98–99, 123
  costing process, 115–123
    approaches to, 115–116
  price, 105–106, 127
  in system life-cycle, 32–34
Warranty analysis process, 105–107
  cost-effectiveness analysis, 106–107
  implementation, 107
  negotiation, 106
  quote, 105–106
  should cost, 105–106
Warranty, checklists, 327–335
Warranty Guidebook, 191
Warranty policies, 16–21
  combination free replacement and pro rata, 17
  defect-free, 19
  failure-free, 18
  fleet, 18
  free replacement, 16
  performance, 20–21
  pro rata, 17
  systemic, 19
  threshold, 18–19
Warranty Review Board (WRB), 84–86, 182–184
Weibull distribution, 29–32
Win-win, 4–5, 41, 105, 135–137

ABOUT THE AUTHOR

James R. Brennan, a well-known warranty practitioner, is principal consultant for Product Assurance Analysts, a Dallas-based consulting business specializing in reliability, maintainability, supportability, quality, life-cycle cost, cost effectiveness, and warranty training and consulting. He has 30 years of industry experience, the last 18 years as a warranty specialist and logistics manager at Texas Instruments, Inc., in Dallas. He conducts numerous life-cycle cost, logistics, and warranty short courses and seminars nationwide for military and commercial customers and suppliers.

Mr. Brennan has written several life-cycle cost and warranty papers, and is a contributing editor to the Society of Automotive Engineers' (SAE) *Reliability, Maintainability, and Supportability Guidebook.* In addition, Mr. Brennan was industry co-chair of the Joint Army–Industry Warranty Requirements Determination Working Group, and is currently conducting related warranty training for Army and contractor personnel. Jim Brennan can be reached at (214) 985-1215 or by fax at (214) 985-0144.